THE USE OF FUNGI AS FOOD AND IN FOOD PROCESSING

author:

WILLIAM D. GRAY

Department of Biological Sciences

Northern Illinois University

DeKalb, Illinois

published by:

CRC Press
Taylor & Francis Group
Boca Raton London New York

CRC Press is an imprint of the
Taylor & Francis Group, an informa business

A DIVISION OF

THE CHEMICAL RUBBER CO.

18901 Cranwood Parkway • Cleveland, Ohio 44128

First published 1970 by CRC Press
Taylor & Francis Group
6000 Broken Sound Parkway NW, Suite 300
Boca Raton, FL 33487-2742

Reissued 2018 by CRC Press

A Library of Congress record exists under LC control number: 70576630

ISBN 13: 978-1-315-89837-7 (hbk)
ISBN 13: 978-1-351-07747-7 (ebk)

Visit the Taylor & Francis Web site at http://www.taylorandfrancis.com and the
CRC Press Web site at http://www.crcpress.com

CRC MONOSCIENCE SERIES

The primary objective of the CRC Monoscience Series is to provide reference works, each of which represents an authoritative and comprehensive summary of the "state-of-the-art" of a single well-defined scientific subject.

Among the criteria utilized for the selection of the subject are: (1) timeliness; (2) significant recent work within the area of the subject; and (3) recognized need of the scientific community for a critical synthesis and summary of the "state-of-the-art."

The value and authenticity of the contents are assured by utilizing the following carefully structured procedure to produce the final manuscript:

1. The topic is selected and defined by an editor and advisory board, each of whom is a recognized expert in the discipline.
2. The author, appointed by the editor, is an outstanding authority on the particular topic which is the subject of the publication.
3. The author, utilizing his expertise within the specialized field, selects for critical review the most significant papers of recent publication and provides a synthesis and summary of the "state-of-the-art."
4. The author's manuscript is critically reviewed by a referee who is acknowledged to be equal in expertise in the specialty which is the subject of the work.
5. The editor is charged with the responsibility for final review and approval of the manuscript.

In establishing this new CRC Monoscience Series, CRC has the additional objective of attacking the high cost of publishing in general, and scientific publishing in particular. By confining the contents of each book to an *in-depth treatment* of a relatively narrow and well-defined subject, the physical size of the book, itself, permits a pricing policy substantially below current levels for scientific publishing.

Although well-known as a publisher, CRC now prefers to identify its function in this area as the management and distribution of scientific information, utilizing a variety of formats and media ranging from the conventional printed page to computerized data bases. Within the scope of this framework, the CRC Monoscience Series represents a significant element in the total CRC scientific information service.

B. J. Starkoff, President
THE CHEMICAL RUBBER Co.

This book originally appeared as part of an article in *CRC Critical Reviews in Food Technology*, a quarterly journal published by The Chemical Rubber Co. We would like to acknowledge the editorial assistance received by the Journal's editor, Thomas E. Furia, Geigy Industrial Chemicals. Mr. C. W. Hesseltine, U. S. Department of Agriculture, served as referee for this article.

AUTHOR'S INTRODUCTION

During more than three decades of teaching mycology (with heavy emphasis on the applied and physiological aspects) the writer has wished many times for a single work describing all of the ways in which fungi are connected in a beneficial manner with food production. The need for such a work has become especially poignant in the last ten or fifteen years during which time it has become increasingly apparent that the rate of human population increase is rapidly pushing the inhabitants of this world into attempts to solve the very real and serious problem of providing sufficient quantities of protein food for all members of the human species. Unfortunately, no such work existed: and, as in many such instances, the choice had to be made between doing it oneself or doing without.

The present work is an attempt to bring together in some sort of organized form all such information that would link mycology (other than the involvement of fungi in food spoilage) to the food industry. It may be justly criticized for its brevity and, in some instances, will probably be criticized for the philosophy expressed. For this the writer makes no apologies. In the first instance, the present discussion is by no means intended to be an exhaustive treatment of the subject. On the contrary, if it serves in some small measure to alert the student to the vast potential resident in fungi, its purpose will have been served. In the second instance, one does not develop and live with a philosophy of mycology most of his adult life merely to change it to avoid offending the occasional sensitive reader. Therefore, in this discussion a spade is not delicately referred to as a "geotome" – on the contrary, if the opportunity exists, a spade is more realistically referred to as a "bloody big shovel."

The section entitled Oriental Fungus-Fermented Foods should at this time be considered merely an outline. It is only in recent years that these curious and, in some instances, delicious food items have received much attention in the Occident; hence, the number of articles printed about them in English is relatively small. However, it is safe to predict that as time goes on we may anticipate the appearance of a much greater number of important research reports on this subject. The relatively recent work of Hesseltine and his associates at the Northern Regional Research Laboratory, Stanton (now at the University of Malaya) and Wallbridge at the Tropical Products Institute in London, as well as the work of others, is serving to focus attention on the fact that Oriental peoples may know some important things about the use of fungi in food processing which are largely unfamiliar to Occidental peoples.

Whether or not to include a discussion of the toxic properties of certain fungi resulted in considerable self-debate. However, it was finally decided that to ignore this unpleasant aspect of fungi would be unrealistic. It must be stated here that the evidence to date indicates that most fungi are not toxic; however, it would be the better part of good judgment to avoid those which are.

William D. Gray
DeKalb, Ill.

THE AUTHOR

William D. Gray is a professor in the Department of Biological Sciences, Northern Illinois University, DeKalb, Ill., having recently been at Southern Illinois University.

Dr. Gray acquired his A.B. degree from DePauw University in 1933 where he received the Distinguished Alumnus Award in 1965. His Ph.D. degree was awarded by the University of Pennsylvania in 1938 when he was also named a National Research Council Fellow at the University of Wisconsin.

Dr. Gray has authored three books and more than 60 research reports in various journals.

TABLE OF CONTENTS

INTRODUCTION

For many years man has been engaged in an endless struggle for survival, not the least facet of which has been his striving to obtain sufficient food. In the course of the many millenia of man's existence as a species on this earth he undoubtedly tried eating many edible appearing objects and finally settled on a relatively small number of other living species (plant and animal) as his principal sources of food. When the pitifully small number (in terms of total species) of plant and animal species which man has used as food is viewed, it becomes apparent that either the number of edible species is actually quite small or else man has demonstrated a remarkable lack of initiative in his search for food. In spite of the fact that there are many thousands of untested (as food) microorganisms, man has used only a few as food and for the most part, at least in the Occident, these are types which at some stage of their life cycle produce a structure which is large enough for him to stumble over. Not possessing any aids to vision, it is understandable why early man largely overlooked microorganisms as a food source (although he certainly learned very early how to use microorganisms in the preparation of a better-textured bread as well as products which were charmingly known to the ancients as "wine and strong drink"). Man has now been in possession of the microscope for several centuries, and biological sciences have advanced to the point where an amazing store of knowledge has accumulated concerning microorganisms —in virtually every area except as regards their potential as food sources. There is no longer any valid excuse for this appalling lack of knowledge, especially in view of the fact that the human species is now facing a major crisis because of the rapidly increasing population. This is a slight misstatement, because it can scarcely be viewed as a problem of the future when protein malnutrition already exists in two-thirds of the populated areas of the earth. At the risk of calling down the wrath of many of the world's biologists on his head, this reviewer wishes to suggest that this world conceivably could be made a much more comfortable and peaceful place to live if at least fifty per cent of all biologists would devote the major part of their efforts to solving the ever-increasing food problem rather than titillate their own imaginations and egos with trivial investigations of one step in a series of enzyme-catalyzed reactions which often lead to a dead end in metabolism, or in the preparation of lists of flora and fauna from an area so small and insignificant that one would be hard put to judge it as adequate for use for any purpose other than as a medium sized building lot.

The present review is an attempt to describe how filamentous fungi* have been used in the past and are used in the present in various areas of the world in connection with man's eternal attempt to satisfy his urge for food. In addition, especial attention has been given to the potential which such fungi seem to possess for making large contributions to the world protein supply. Although mushrooms of one type or another have been eaten by man for many years, their contribution to world food supply must be considered negligible, simply because in most instances they are used as condiments rather than as food staples. In spite of the fact that truffles are often viewed as the gourmet's delight, their actual contribution to the world's food supply has been and will remain infinitesimally small because the stark facts are that only a small fraction of a per cent of the people on this earth are gourmets, and hence only a few ever taste these gourmet items. Yeast will not be considered in the present account for two reasons: (1) they are highly atypical fungi which ordinarily produce no true mycelium, and (2) so much fine research has been conducted with yeast that they deserve to be the sole subject of a separate account.

The major past and present contributions of fungi to man's food supply must be sought in the Orient; but as the Occident encounters the population pressures already encountered many years ago in the Orient, it may be that a more rational attitude toward fungi will emerge. At this time it must be noted that a curious ambivalence exists in the human race with respect to fungi. In the Orient several mold-type fungi have been used for centuries in the preparation of a variety of food products, but in the Occident this use has been restricted to the preparation of two types of cheese. The extreme Occidental bias against fungi reaches its real climax of absurdity in the beverage alcohol regulatory laws of the United States: For years it has been known that grain starches can be more economically converted to fermentable sugar by mold enzymes than by malt enzymes; however, whiskey is so defined by law in the United States that it cannot be legally made using mold enzymes. The average Occidental has no qualms about receiving a dose of penicillin, a medicine that is synthesized by a fungus. Also, he has no reservations about drinking a soft drink, most of which contain an organic acid synthesized on a commercial scale through use of a fungus. On the other hand, if one suggests to the average citizen that he eat some fungus other than a mushroom, immediate resistance to the idea is usually encountered. Gray[114] has attributed this widespread Occidental antipathy toward fungi in part to the effects of Biblical and classical literature; however, this is probably an oversimplification, and some of the antipathy of the layman can probably be attributed to the fact that fungi are often especially abundant in moist, dark places where there is an abundance of decaying organic matter—places viewed as quite unpleasant by many people. Curiously enough the very people who would not ever consider eating a food on which a mold has grown are the very people who would not consider eating a steak unless it was well-garnished with mushrooms which obtained the major part of their sustenance from horse manure. These same people are often those who are inclined to eat such bizarre items as rattlesnake paste, chocolate coated ants, salted cherry blossoms and French-fried agave worms!

The present review is presented in several sections, the principal ones being concerned with (1) the ways in which fungi have been and are still being used directly as food, (2) the ways in which fungi have been used in the processing of various food products either for the purpose of making a different, a more palatable, or a more nutritious food, and (3) the ways in which, in this writer's opinion, fungi might well be used to make significant contri-

*The term filamentous fungus is here applied to any fungus which develops a true mycelium.

butions to the world protein supply and thus help in the solution of the major problem facing mankind today. The final major section is concerned with a consideration of fungus toxins, since it would be foolhardy to attempt to create the false impression that there are no toxic fungi. Some very real toxins are to be encountered among the fungi, and any account that purports to discuss fungi as food must take these into consideration. In the writer's opinion the evidence to date does not indicate that a very large proportion of the fungi is toxic, and he here reaffirms the statement which he first made over a decade ago, that when the beneficial effects of fungi are weighed against the harmful effects, the overwhelming balance is on the side of the beneficial. It should be emphasized that this and other opinions are the writer's alone; and if they are incorrect, he must face the criticism alone.

The production of vitamins or fats, per se, by filamentous fungi is not discussed since they are considered food additives and the present review is concerned with foodstuffs as such. In no instance has an attempt been made to cite every paper that has been written on a particular subject. It would be impossible to do this in a review such as the present one, since reports on production of the Common Cultivated Mushroom now number many hundreds as do reports on a single fungus toxin, aflatoxin. On the contrary, an attempt has been made in each instance to present as briefly as possible those reports which contribute something of concrete nature to the development of a broad picture of the particular subject under discussion. Another reviewer might well depend upon quite different reports to present the picture. In any event, the reader who is interested in some particular phase will find that the bibliographies of the papers cited will lead him further and further into the more specific details of that phase.

DIRECT USE OF FUNGI AS FOOD

Mushrooms as Human Food

With some mycologists the use of the term "mushroom" is restricted to the umbrella-shaped reproductive structure (carpophore or basidiocarp) with gills on the lower surface which is produced by a member of a specific group of the class Basidiomycetes. The laymen's concept of a mushroom is often somewhat different since it would include the edible reproductive structures of any fleshy fungus regardless of whether or not it is an agaric. There are some fleshy fungi which are known to be poisonous, and the mycologist would ordinarily simply refer to these as poisonous mushrooms. On the contrary, the layman quite frequently refers to poisonous mushrooms as "toadstools." The term toadstool is regarded here as being so meaningless that it will not be used in the present discussion. Here the term mushroom will be used with reference to all fungi which have umbrella-shaped carpophores with gills (edible or non-edible, poisonous or non-poisonous) plus all other fungus carpophores (regardless of the class to which they belong) which are fleshy and edible. Thus, in the present sense, edible truffles (which are not agarics) will be considered as mushrooms, whereas inedible ones will not be so considered.

How long man has been eating mushrooms is, of course, impossible to determine, but one can speculate with reasonable assurance that such fungi have periodically (because of their seasonal occurrence) been a part of his diet for many centuries. Chavez[56] has suggested that man has been familiar with beverage alcohol for at least three hundred centuries, and it is suggested here that mushrooms have been a part of the human diet for an even longer period of time. The reason for this suggestion is a simple one: discovery of the alcoholic fermentation undoubtedly occurred only after an episode of forgetfulness on the part of a man, whereas the discovery that some fleshy fungi are edible merely depended upon some hungry or curious person finding and eating them. Furthermore, primitive man probably would not hesitate to test the edibility of such an object, since he would not have been influenced by the many centuries of superstition and folklore that seem to have developed in connection with fungi. Thus, there seems no reason to doubt that during the hunting and gathering stage of his development man tried an occasional mushroom when he found it, since apparently there were times when he was hard put to find sufficient food and probably tried eating just about anything that appeared edible. Unquestionably it was by this system of trial and error that the human species accumulated a body of knowledge concerning those fungi which are edible and those which are not. Certainly before the 12th century A.D. it was believed in the Scandinavian countries that the fly agaric, *Amanita muscaria,* was an hallucinogenic fungus and could be used to induce certain symptoms of schizophrenia (Gray[114]).

On the basis of brief references to manna in the Old Testament, especially with regard to the sudden appearance of this material, it is a temptation to suggest the possibility that the children of Israel were eating mushrooms which often spring into being with almost miraculous rapidity at certain seasons of the year. Like various other suggestions regarding the identity of manna, the present one can neither be proved nor disproved. However, the true gourmet, who is so fond of mushrooms, would probably reject this suggestion, since he would note that after a time the children of Israel objected to having manna on their menu so frequently (*Numbers* 11: 5-6).

While our knowledge of mycophagy prior to the time of written history must of necessity remain purely speculative, we are burdened with no such restrictions starting with Greek and Roman times. Classical Roman literature abounds with references to the eating of mushrooms, and these were more often than not viewed as extreme delicacies. At one time during the heyday of the Roman Empire one could easily judge the degree of esteem in which his dinner host held him by the quality (species or mode of preparation?) of the mushrooms that were served to him. The esteem in which the Romans held mushrooms is evidenced by the report and photograph published by Harshberger[134] in which the writer described a mushroom carved in stone in a Roman ruin in North Africa. Furthermore, the Romans were certainly in possession of knowledge concerning both poisonous and non-poisonous mushrooms, since Lucretia Borgia is reputed to have eliminated at least one of the people that she didn't like by feeding him what the modern layman would probably term a "toadstool." Ramsbottom[236] reported that a painting in Herculaneum (79 A.D.) figures a mushroom, and this same author has reproduced a fresco from an old French church (1291 A.D.) in which the fly agaric (*Amanita muscaria*) is shown. Ramsbottom interprets this latter illustration as a representation of this mushroom as the tree of "good and evil." If this interpretation is correct then there may have been an existent belief that this species was both hallucinogenic and lethal. Wasson[328] denies that *Amanita muscaria* has such properties on the basis of his having eaten some. However, Wilkinson[334] has reported that the active principle of *Amanita muscaria* is muscarine (Figure 1), which may be rather significant, since Eugster and Muller[91] have found muscarine in species of *Inocybe,* some of which are known to be hallucinogenic.

FIGURE 1

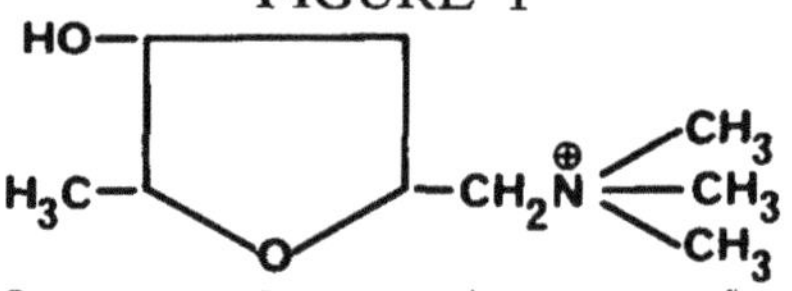

Structure of muscarine, the active principle of *Amanita muscaria* (the fly agaric) and some species of *Inocybe.*

Following the example set by the people of Roman times, the eating of a wide variety of mushrooms collected in the wild has persisted in Europe and other areas of the world up to present times (Figure 2), a rather limited, although certainly not unknown, practice in the United States. However, in this country a relatively small percentage of people are even aware of and, hence, do not take advantage of the occurrence of the delicious morels, and an even smaller percentage collect and eat a variety of other edible fleshy fungi. Gilbert and Robinson[108] estimate that less than one person in a thousand in North America recognizes any species of mushroom, and the frequently-published humorous cartoons of the popular magazines probably well-illustrate the extent of the knowledge of mushrooms possessed by the vast majority of American citizens. Thus, aside

FIGURE 2

Irish farm boy selling the field mushroom (*Agaricus campestris*) which he has collected. Main road to Limerick; near Kilcolgan, Ireland, September 1969.

FIGURE 3

Dried mushrooms being offered for sale in the African market of Harari Township. Salisbury, Southern Rhodesia.

from an occasional box of morels, this writer has not seen a fresh, wild mushroom offered for sale in a domestic market for at least three decades. Quite a different situation exists in many parts of Europe where, in season, a wide variety of edible species is presented for sale in the markets. This practice is not exercised wholly without controls because Rolfe and Rolfe[244] note that as early as 1897 only certain mushroom species could be offered for sale in the market of St. Etienne. That mycological knowledge expanded during the ensuing two decades is evidenced by the fact that the regulations were altered in 1921. Singer [265] states that wild mushrooms are one of the most valued vegetable crops of the Arctic, especially the arctic or subarctic regions of Europe and Asia. They are eaten fresh or are pickled or dried for use in winter.

A variety of mushrooms of one type or another is collected and marketed in the Orient. Not only is Shiitake cultivated in Japan, but naturally-occurring carpophores of this mushroom are collected from their natural habitats. Another highly-prized mushroom in Japan is Matsutake (*Tricholoma matsutake*—pine mushroom—Japanese mushroom). *Tricholoma matsutake* forms an association (mycorrhiza) with pine roots and, hence, is to be sought in pine woods where it may be collected for eating fresh or prepared for the market by drying. Still another fungus used in the Orient (but especially in China) is the Ear Fungus (*Auricularia polytricha*) which may be cultivated by primitive methods but is commonly collected in the wild and dried for market.

There are no published data on the extent to which wild mushrooms are used by the native populations of subequatorial Africa. In only one instance has the writer seen dried mushrooms offered for sale in a native market (Figure 3). Two different species were offered for sale in the market in Harari Township, Salisbury, Southern Rhodesia, but they were in poor condition and no attempt was made to identify them. With no great abundance of food in certain areas of Africa, mushrooms might make a very welcome addition to the diet, and it seems quite possible that mushrooms are eaten there rather more frequently than the writer's rather casual observations revealed.

It seems especially unfortunate that there is not a greater general awareness among laymen, especially in the United States, that there is a rather large number of edible mushrooms to be found in season in their natural habitats in this country. Of the several thousand species of fleshy fungi which occur here,

FIGURE 4

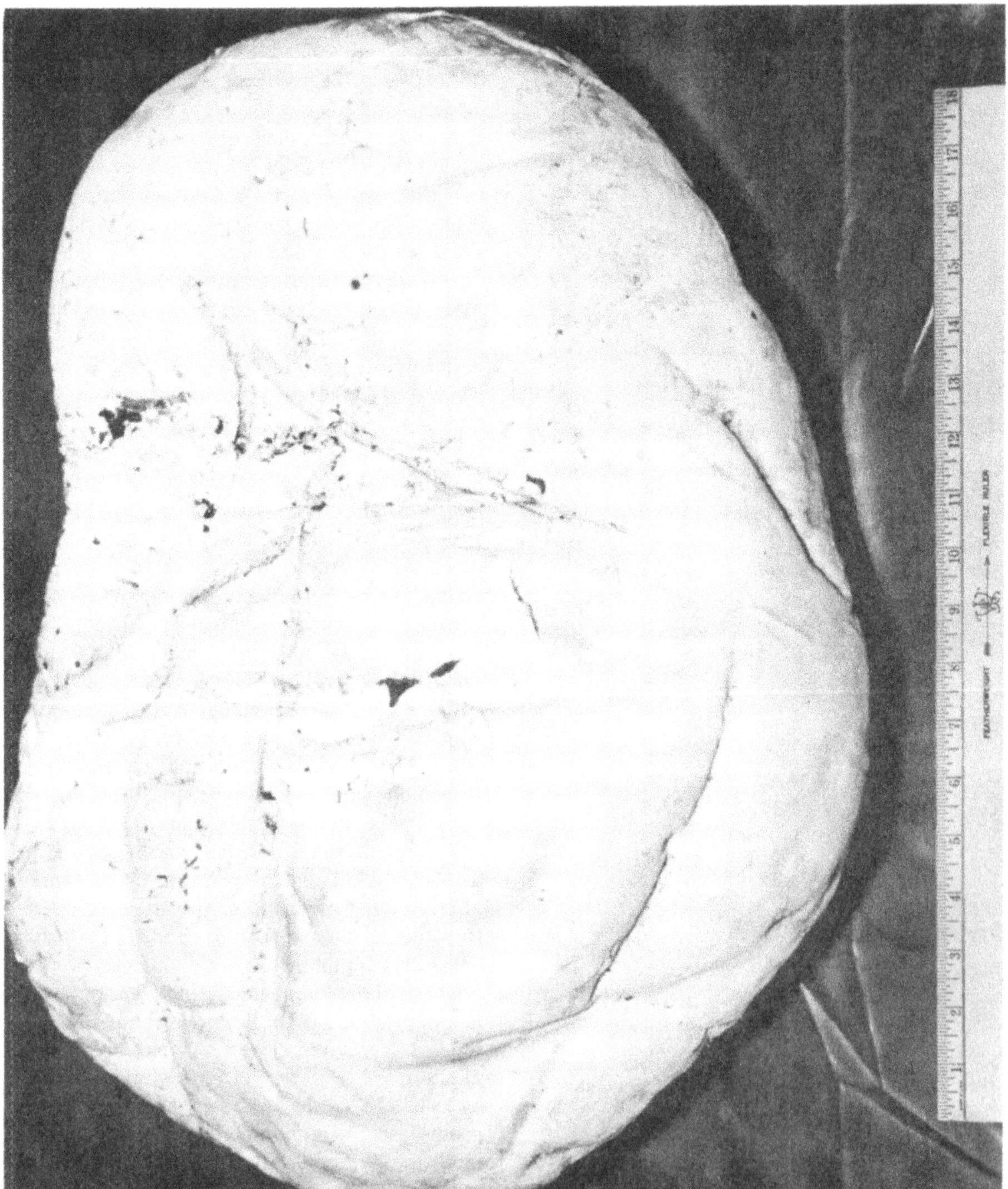

The giant puffball (*Calvatia gigantea*). Length—21 inches, width—14 inches, height—14 inches, weight—16 pounds, 7 ounces. From *The Union Forever,* 16, No. 10, 4, October 1957. Courtesy of the Union Fork and Hoe Co.

an extremely low percentage of them is represented by poisonous types (Smith[272]) and many are better flavored than the commercially grown mushroom. Five different types of mycetismus (mushroom poisoning) have been described. For the reader who is interested in detailed descriptions of symptoms, the early works of Ford[97,98] should be consulted.

It is certainly not suggested here that wild mushrooms be collected indiscriminately and eaten with impunity by the uninitiated, but it is suggested that it is not especially difficult for the average person to learn to quickly and accurately identify ten or a dozen edible wild species. For purposes of identification manuals such as those of Christensen[58] or Smith[273] may be used. Among the easier types to learn would be various species of "puffball." No puffball species of the genera *Lycoperdon* or *Calvatia* are known to be poisonous, although none would be palatable if not fresh. Therefore, any species in this group can be eaten with safety if it is fresh. Freshness can be established by breaking the puffball open; if the tissue within is firm, snow-white in color and has the texture of moist angel food cake, the puffball is fresh. Many different species of puffball occur in the United States. They range in size (depending on species) from small, spherical or pear-shaped carpophores less than an inch in diameter to the occasionally encountered carpophores of the giant puffball (Figure 4), which may sometimes attain a diameter of over two feet. Works such as those of Christensen[58] or Coker and Couch[64] as well as others may be used for the identification of the common puffballs, which are most abundant in the late summer and early autumn.

Of those laymen who collect and eat naturally-occurring mushrooms, the greatest number seem to be familiar with various species of *Morchella* (morel). All species of morel appear in the early spring and have fleshy fruiting structures not at all suggestive of the umbrella-shaped basidiocarp of the cultivated mushroom. Several species occur in the United States; but, in the writer's experience, laymen who are familiar with the morel most frequently collect four species: *Morchella esculenta* (sponge mushroom), *Morchella hybrida* (spring mushroom), *Morchella crassipes* (thick-footed morel) and *Morchella angusticeps* (black morel), all of which are shown in Figure 5. There is one species, (*Gyromitra esculenta*—false morel), not one of the morels but closely related to them, which is poisonous. Only extreme ignorance or carelessness would result in the false morel being confused with a true morel, since *Gyromitra esculenta* typically is much larger and bulkier than any true morel (the stalk is rugose and may have a diameter of over three inches in large specimens), and the upper part (pileus) has a dark brown, sometimes reddish color not seen in any true morel. For the beginning morel hunter it is advised that only morels that look exactly like those in Figure 5 be eaten until a thorough knowledge of the group is obtained.

Considering the fact that there is good reason to believe that man has been eating naturally-occurring mushrooms for many centuries and the additional fact that there are several thousand species of such fungi, many of which are quite tasty, man has demonstrated a remarkable lack of initiative in the deliberate cultivation of mushrooms. In his extensive and excellent work *Mushrooms and Truffles,* Singer[266] lists only four species which have been produced to such an extent that they require much consideration here. If one relies on the older literature it would appear that a far greater number of species have been cultivated by man, but this greater number is more apparent than real because there has been considerable taxonomic confusion regarding the identity of the cultivated forms. Thus, one may find the Common Cultivated Mushroom of the United States and Europe referred to as *Agaricus campestris, Psalliota campestris* and *Agaricus bisporus.* Apparently *Agaricus bisporus* is the only one which has ever been cultivated, *Agaricus campestris* being an altogether different species which occurs only in the wild state, and *Psalliota campestris* being a binomial which is invalid according to generally accepted rules of nomenclature.

The four cultivated species recognized by Singer are *Agaricus bisporus* (the Common Cultivated Mushroom of the Occident which is still frequently incorrectly called *Agaricus campestris*), *Volvariella volvacea* (the Padi Straw Mushroom), *Lentinus edodes* (the Shi-

FIGURE 5

Common edible species of morel: A. *Morchella angusticeps,* the black morel; B. *Morchella crassipes,* the thick-footed morel; C. *Morchella esculenta,* the sponge mushroom; D. *Morchella hybrida,* the spring mushroom.

FIGURE 6

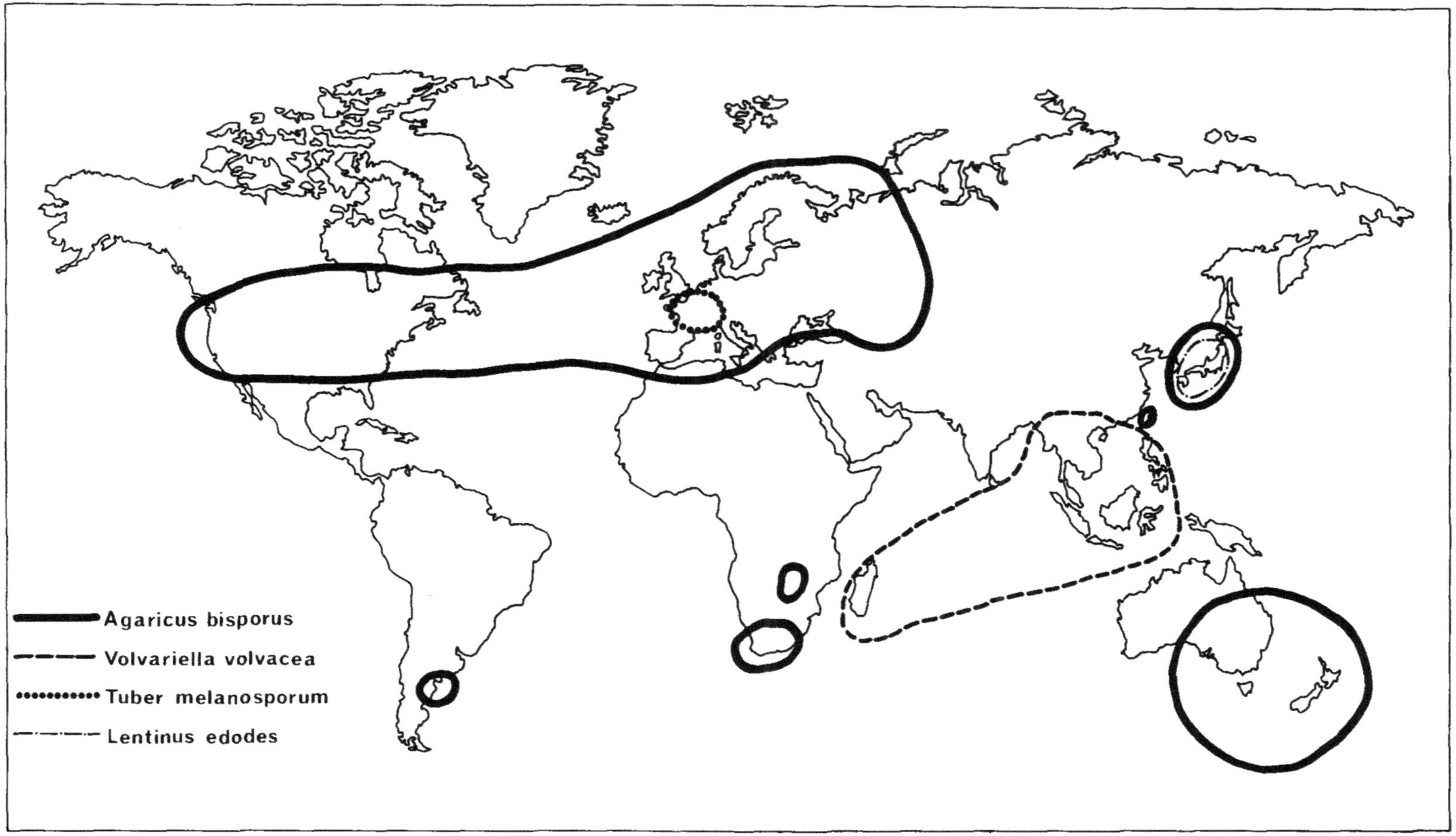

Principal areas of cultivation of *Agaricus bisporus* (Common Cultivated Mushroom), *Tuber melanosporum* (Truffles), *Volvariella volvacea* (Padi Straw Mushroom), and *Lentinus edodes* (Shiitake).

itake of eastern Asia) and *Tuber melanospermum* (the Perigord Truffle). Truffles grow in association with oak roots, the Padi Straw Mushroom on moistened straw, Shiitake is grown on small logs, and the Common Cultivated Mushroom is grown on specially composted horse manure. The principal areas of cultivation of these four mushrooms are indicated in Figure 6. Since man's participation in the cultivation of the truffle is a rather indirect one and involves less control and fewer unit operations than cultivation of any of the other three, this type of mushroom production will be discussed first.

Cultivation of Truffles

The person who customarily eats only the Common Cultivated Mushroom would in all probability not recognize the truffle as a mushroom. In the first place, truffles are dark in color, warty in appearance and tuberous in shape, and in the second place they grow only underground. Truffles do not even belong to the same class of fungi as the Common Cultivated Mushroom, since they are Ascomycetes not Basidiomycetes. Nevertheless they meet the requirements of a broad definition of mushrooms in that they are edible fleshy fungi and hence will be termed mushrooms in the present discussion. There are several species of truffles (members of the genera *Tuber* and *Terfezia*), and a number of black truffles are collected in the wild state and marketed. However, deliberate attempts to cultivate truffles have been made in connection with only one species, *Tuber melanospermum* (the Perigord Truffle) and so the present discussion will be devoted primarily to this species.

These curious and delicious fungi apparently grow only in association with certain tree roots, a fact which Singer[266] states was discovered in 1810 by a man named Joseph Talon. Thus, of the mushrooms purposely cultivated for food purposes by man, it is only in connection with the truffle that we have exact information regarding the inventor and the date of the invention. Talon planted acorns on poor soil apparently with the intent of obtaining an oak plantation and found that he could harvest truffles from under the trees a few years later. This observation was later verified by others and since that time many oak plantations have been established in France with the express purpose of providing suitable habitats for the growth of truffles. That truffles grow in association with tree roots is not at all a unique situation among the fungi, since many fungus species are known to form an intimate association with the roots of higher plants, and there is a considerable body of evidence (e.g., Kelley[165]) to support the view that in all probability many woody plants could not even exist in most habitats if their roots were not associated with the proper fungus or fungi. *Tuber melanospermum* has been grown in pure culture in the laboratory by various workers. The present writer has had it in culture for over a decade and others (e.g., Matruchot[206]) cultured it many years ago. However, the reproductive structure (ascocarp), the truffle of commerce, is formed only in the soil in connection with tree roots. Singer[266] notes that claims have been made that truffles grow under at least eight different genera of trees, but that all commercial growers of this mushroom use one of four species of *Quercus* (oak).

Thus, if one wishes to produce truffles commercially he must first plant acorns and obtain an oak plantation. Apparently in France *Tuber melanospermum* need not be introduced into the plantation but will invade naturally, although Malençon[201] has suggested a method by which the introduction can be made. It could be argued that production of the truffle is not mushroom culture at all because other than to select the proper site and to plant the proper species of oak, man makes no further contribution other than to occasionally prune the trees. Nonetheless, since these gourmet items are more frequently than not recovered from oak plantations which were initially established to provide a habitat for such fungi, the practice is considered here as a form of mushroom culture. In his 1938 paper, Malençon[201] distinguishes between direct and indirect methods of truffle production, but in this writer's view such distinction seems pointless.

Truffles vary in size from that of a walnut to that of a fist. They are often described as the most delicious of mushrooms, a description with which many Europeans (and especially the French who sell these mushrooms) would probably agree. However, there are many in the United States who would challenge this description, since they feel that the morel provides the ultimate in mushroom flavor. Powerful arguments can be mounted on either side but in dealing with gourmet items, individual choice still has to be the major consideration. Nonetheless, truffles are highly prized and do command very high prices in the markets of the world. They are produced primarily in France, and Chatin[55] stated that 3.3 million pounds (both cultivated and wild) were sold in 1868, while 4.4 million pounds were sold in 1890. Singer[266] believes these earlier figures to be exaggerated and he cited official production figures for 1935 as being 3,577,200 pounds. An anonymous article in *Newsweek*[10] estimated production in the 1968-69 (November to February) season as being two million pounds and noted that this was double the production of the previous season. Since in early 1969 truffles were priced at $32 per pound in Paris and $41 per pound in New York, it is obvious that the successful truffle grower may derive a very sizeable income from this practice. Not only that but he will have a crop of oak trees to sell when truffles are no longer produced in abundance in a grove. Once the plantation is established six to ten years may elapse before many truffles are produced, but apparently truffle production may then continue in the same grove for many years.

Perhaps one of the attractions of truffles is the fact that they are claimed to be mild aphrodisiacs. The reviewer is unaware of any experimentation that would support or deny this claim. It may simply be part of the folklore of truffles, and since at one time or another a great many substances have been claimed to have aphrodisiacal properties, it is suggested that truffles may be aphrodisiacs to about the same degree as the powdered rhinoceros horn so prized in certain parts of the Orient or the carved rhinoceros horn cups so treasured in various parts of the Arabic world.

A considerable body of folklore has grown up in connection with the harvesting of truffles. Because of their subterranean habit, the finding of truffles is often believed to present something of a problem, since, while the harvester knows that they occur in the soil in oak groves, it is thought that he doesn't know exactly where to dig. Thus, young dogs and pigs were trained not only to detect truffles by scent but also to dig them out for their masters. A number of humorous stories and beliefs have developed around this practice, but the present writer is totally unprepared to accept the view that young female pigs in their first pregnancy are necessarily the best truffle hunters. Lohwag[196] states that a skilled human truffle hunter does not need the help of either dogs or pigs but can find these underground fungi by searching for such indications of their presence as the color of the surface vegetation, convex patches of earth with small cracks, etc. Singer states that the use of trained dogs and pigs in truffle havesting is not as generally practiced as the layman believes. In spite of these facts, the concept that truffle hunters require canine or porcine assistants is still fostered in the lay mind by such articles as those that appear in *Newsweek* (February 24, 1969, pp. 72, 77) and the *National Geographic* (Volume CX, No. 3, 1964, pp. 419-426).

The present discussion of truffle cultivation is of necessity quite brief (Chatin devoted 202 pages to it in 1892) and was designed to acquaint the reader only with the major points of the practice. Many details have been omitted. For example, selection of soil type, selection of site and selection of proper species of oak are quite important and such selections may vary from area to area in France. Nothing has been said about the manner in which most growers prune their trees and that, as in the production of all agricultural crops, the growers have good seasons and they have bad seasons. There are pests of truffles and there are probably diseases of truffles. The complete life history of *Tuber melanospermum* is still not clearly understood. For a more detailed description of the truffle and its cultivation, the works of Chatin[55], Malençon[201] and Singer[266] are recommended.

Cultivation of the Common Mushroom

Of the four mushrooms which are cultivated in quantity by man, by far the greatest amount of research and development has been conducted in connection with the Common Cultivated Mushroom, *Agaricus bisporus*. According to Singer[266] the cultivation of this type of mushroom began in about 1700 with its introduction by French horticulturalists. While the French introduced present methods of growing mushrooms in caves and mushroom houses, and American mushroom growers refined and elevated house culture almost to factory status, Singer gives credit to the Swedes for having developed the greenhouse culture of such fungi. Since modern mushroom production methods are most highly developed in the United States, the present discussion will deal primarily with methods currently in use in this country.

Stages in the commercial production of *Agaricus bisporus* are eight in number. They are as follows: (1) spore germination, (2) spawn growing, (3) composting, (4) filling and sweating out, (5) spawning, (6) casing, (7) the cropping period and (8) packing and marketing. The grower usually does not perform Steps 1 and 2 but obtains his pure culture spawn from a commercial spawn laboratory of which there are a number of reliable ones in this country. With the exception of Stage 3, the steps are almost as closely controlled as production steps in any modern factory.

The process of spore germination as a step in mushroom production dates only from the end of the nineteenth century, although mushrooms had at that time been cultivated for about two centuries. Astonishing as it may seem to anyone who has observed how readily and rapidly most fungus spores germinate, it was not until about 1890 that two French mycologists, Costantin and Matruchot, solved the problem of spore germination in *Agaricus bisporus* and thus made the production of pure culture spawn possible. Their technique remained a closely guarded secret, and for about two decades the Pasteur Institute enjoyed a virtual monopoly in supplying pure culture spawn grown from spores. One wonders what the great Louis Pasteur's reaction might have been to such secrecy in the institute bearing his name in view of his own public demonstration of an anthrax immunization procedure! This spirit of secrecy, shrouded in mysterious activities, undoubtedly made its contribution to the present day reticence toward fungi so evident in such a large percentage of the population, although it was undoubtedly quite profitable for the Pasteur Institute for a number of years. The method of obtaining spore germination was rediscovered in the United States by Ferguson,[94] but for a number of years the method of Duggar[85] was used for the production of pure culture spawn in the United States. This latter method, although it does not involve spore germination, deserves mention here since it really marks the beginning of more scientific mushroom cultivation in the United States. Duggar's method is a tissue culture method and can be quite easily performed with very little practice even by a novice. It merely involves breaking a fresh, fleshy, leathery or gelatinous carpophore and quickly transferring a small piece of the inner tissue from the freshly-broken surface to a suitable sterile medium with a sterile transfer needle. Many mycologists have used this technique, and the writer has successfully employed it for the past three decades to obtain a variety of fleshy fungi in pure culture. However, today most pure cultures of *Agaricus bisporus* are obtained by germination of spores since spawn so derived seems superior and it also offers a better opportunity to select different and perhaps better strains. Spawn can be obtained from spores by a variety of methods, but the method of Stoller and Stauffer[289] seems quite adequate. For starting a pure culture, spores from a perfect specimen of a good crop of the Snow White variety of *Agaricus bisporus* are generally used in this country. In Europe the darker variety of *Agaricus bisporus* is preferred in some areas.

Once the mushroom has been obtained in pure culture from spores, the master culture (or sub-cultures from it) is used to inoculate large quantities of proper substrate in order to prepare pure culture spawn. Prior to the development of pure culture spawn, mushroom beds were inoculated with spawn from other mushroom beds; but use of such spawn often caused problems. Since it had been grown on non-sterile substrate, contaminants had a tendency to build up in number and variety, and also the mushroom mycelium often lost vitality. In such instances the grower said that the spawn had "run out". Hence, only pure culture spawn is used today in the United States. When first developed, pure culture spawn was prepared by inoculating sterilized, composted horse manure (the same type of compost that would be used later for growing the mushrooms in quantity) in milk bottles. Two other types of media are now often used for the growth of spawn: (1) Grain Spawn, hard winter-rye grain mixed with calcium carbonate and water, and (2) Tobacco Process Spawn (see Rettew[238]), a mixture of tobacco stems with humus and/or peat. Other substrates have been used for spawn making; however, Tobacco Process Spawn seems to be most widely used. Singer[266] expresses the opinion that claims about speed of growth and yield depending upon use of a certain brand of spawn are unjustified. Thus, Langkramer and Rĕzník[187] prepared pure culture spawn on a medium that they used as a compost substitute and observed that the mycelium grew about twice as fast as it did on sterilized, composted manure. As Singer[266] has pointed out, however, this does not necessarily mean that such spawn is superior for commercial use. In all probability, spawn purchased from any of the experienced spawn makers would prove quite satisfactory if properly inoculated into good compost and if suitable conditions for growth were maintained.

Once the substrate (usually tobacco stems) is inoculated, it is incubated at about 21°C and when thoroughly permeated with mycelium, the spawn is ready for use. It may be used fresh (usually within a week after being removed from the incubator room), it may be stored for several months in a cold storage room, or it may be carefully dried and kept viable for over a year in a cool, dry place.

Continuous efforts are being made to select better strains of *Agaricus bisporus*. For example, Kneebone[171] has briefly described the obtaining of over 1500 single spore, multispore, tissue, and mycelial transfer cultures and their subsequent testing. When their productive capacities and quality features have been determined at the Mushroom Research Center at Pennsylvania State University, and selected strains have performed well through at least three successive crops, they are released to the mushroom industry.

Insofar as most commercial mushroom growers are concerned, there has been little change in the substrate on which mushrooms are grown since the process was invented in France in about 1700. Traditionally this substrate is composted horse manure, and today, nearly three centuries after the industry had its beginning, horse manure is still the major substrate constituent in most instances. The early establishment and expansion of the mushroom growing industry in the United States near Kennett Square, Pennsylvania, was not accidental. That region happened to be the area where the many horses (used for drawing the early horse cars of Philadelphia) were stabled, and, hence, an adequate supply of manure for composting was readily available. With the passage of the horse from the American agricultural scene, it was feared that sufficient supplies of horse manure would no longer be available and many attempts were made to devise a synthetic compost (e.g., Hutchison and Richards,[154] Waksman, Tenney and Diehm,[324] Waksman and Reneger,[323] Jenkins,[157] Treschow,[308] Rettew and Thompson,[239] Yoder and Sinden,[347] etc.) but most mushroom growers still make their compost wholly or in large part from horse manure. Fears that the passage of horse-powered agriculture might have a serious effect upon the mushroom industry now seem ill-founded in view of such reports as that from Ohio where in 1963 there was a greater horse population than there had been during the days before gasoline-powered farm equipment was in common use. Nonetheless it is not uncommon to add other materials to manure to "extend" it, especially when the price of manure is high, and the writer has observed good yields of mushrooms obtained in Southern Rhodesia on compost made from manure to which dried, wild veldt grasses had been added (Figure 7).

Before being used as a substrate for mushroom growing, horse manure (no other manure is suitable) must be composted. In this process the manure (plus whatever other organic matter may have been added as an extender) is heaped up in long piles (4 to 5 feet wide, 8 to 16 feet long and 3 to 6 feet high) and allowed to undergo a natural fer-

FIGURE 7

Tray beds with first break of *Agaricus bisporus* just starting. Compost made from horse manure extended with wild veldt grasses. Penhalonga, Southern Rhodesia.

FIGURE 8

Mixtures of horse manure and wild veldt grasses being composted. Penhalonga, Southern Rhodesia.

mentation. Composting is usually done out of doors (Figure 8) or under an open shed and represents the one phase of the mushroom industry least subject to control. The non-uniform character of the processes which occur in a pile of composting manure are well-illustrated in Figure 9. During composting, which ordinarily requires fifteen days or more, the manure is turned and mixed at intervals to obtain a more uniform product. Sinden and Hauser[263] have devised a composting procedure which reduces composting time to ten days or less, and Delmas[75] has described an experimental fermentation container in which composting can be accomplished under controlled conditions. In 1941, Lambert[184] described an "indoor composting method" in which the final stages of the composting process are conducted in an indoor composting-house where temperature and moisture conditions can be controlled. It seems possible that ultimately a method of composting embodying the concepts of these four methods may be developed and thus place the process of composting on a somewhat more controlled and uniform basis. Such a method, of course, must be able to compete economically with presently used composting methods.

In view of the chemical complexity of a naturally-occurring material such as horse manure, it is understandable that no one knows exactly what happens during composting, a process in which manure undergoes a complicated natural fermentation which is brought about by the many different microorganisms that normally occur in manure. The objective of composting (Lambert[182]) is to prepare a substrate that is more suitable for the growth of *Agaricus bisporus* mycelium than for the many microorganisms whose presence in a non-sterile material such as manure cannot be avoided. During composting, easily decomposed materials, such as sugars, starches and hemicelluloses disappear at a much more rapid rate than cellulose and lignin, and at the same time insoluble nitrogen compounds accumulate, probably as a result of the assimilatory processes of the microorganisms in the manure. Styer[290] and Waksman[321] have demonstrated that lignin and insoluble proteins can be utilized easily by the mushroom mycelium. Hence, in a properly prepared compost, the desired fungus can grow much more rapidly than fungi which cannot use these materials so readily. The chemical changes which occur during composting have been investigated by Hebert and Heim,[136,137] Waksman,[320,321] and Waksman and Nissen,[322] but the practical mushroom grower is more interested in the proper procedure for preparing a compost that will give good results. Such empirical procedures are outlined in detail by Duggar,[84,86] Rettew and Thompson[239] and more recently by Singer.[266] The latter worker has stated that regular pH tests may be used in part as tests for the readiness for use of a compost and can assist the grower in prevention of under- or over-composting. During composting, the pH of the outer layers may go as high as 9.0 and when the pH starts to drop, the composting should be stopped.

After the compost has been properly prepared by present methods, it is taken inside (a cave, abandoned mine, or more commonly in the United States inside a specially constructed mushroom house) where it is placed in trays or shelves, or beds are prepared from it on the ground. Rettew and Thompson[239] have pub-

FIGURE 9

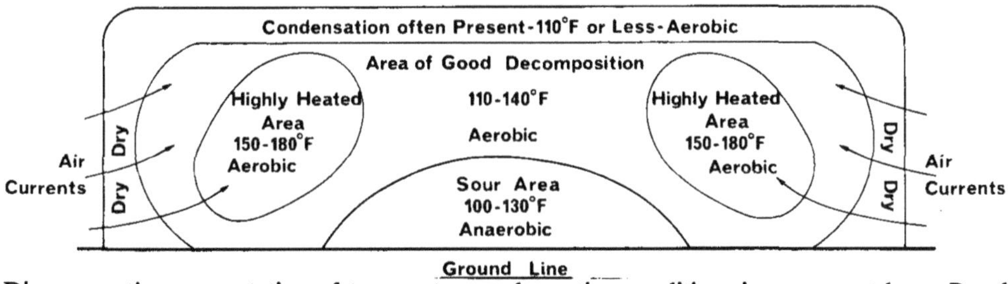

Diagrammatic representation of temperature and aeration conditions in a compost heap. Based upon Rettew and Thompson[239] and Lambert.[184]

lished plans for the construction of a mushroom house and hence details will not be given here. Such houses must be so constructed as to allow for proper heating as well as ventilation, since the accumulation of carbon dioxide must be prevented. As early as 1933 Lambert[181] demonstrated experimentally that a one per cent concentration of carbon dioxide will inhibit basidiocarp development, and a concentration as high as five per cent will completely arrest this development. Hence, adequate means of ventilation must be provided. Although refrigeration is expensive, Singer[266] states that most modern mushroom establishments have refrigeration equipment. Based upon conditions in the Netherlands, Spoelstra[280] has made suggestions regarding the air conditioning of mushroom houses.

Beds may be of three types: (1) tray beds in which the compost is placed to a depth of 4 to 5 inches in individual wooden trays; (2) shelf beds in which compost is placed to a depth of 5 to 8 inches on wooden shelves (with side and end boards) which are in tiers and separated from each other by a vertical distance of 2 to 2.5 feet; and (3) French beds, in which the bed is built directly on the ground to a height of 1 to 1.5 feet. Tray beds (Figure 7) have many advantages and Singer[266] has predicted that they will ultimately dominate in the industry.

When the beds are first prepared the compost is not quite matured, and the "sweating out" period begins. This process in reality is the final more controlled stage of the composting process. During the sweating out period the temperature of the compost rises and should reach a temperature of 130-145°F sometime between the third to fifth day. If the temperature exceeds 145°F the house must be cooled by ventilation; if it does not heat up as high as 130°F the house must be heated. Following heating up, the compost then cools, the entire sweating out process requiring 7 to 10 days. When the temperature is below 100°F but not as low as 80°F the bed must be spawned.

Spawning is the step in which the vigorously growing mycelium which has been received from the spawn laboratory is introduced into the compost. When manure spawn is used, it is broken into pieces about the size of walnuts which are inserted about two inches deep in the compost and then covered without pressure being applied. With Tobacco Process Spawn or Grain Spawn, the spawn is crumbled into small particles and as much as can be taken up between the thumb and two fingers used instead of a nut-sized piece of manure spawn. An uninoculated zone of about four inches is left along the margins of the bed (tray or shelf) and the rest of the bed is inoculated at points which are check-rowed and about eight inches apart. Variations of this procedure may occur. For example, about two inches of compost may be removed from the top of the bed, tobacco spawn broadcast over the surface, and the top layer of compost then replaced. The inoculation of French beds is accomplished somewhat differently, but in all instances inoculation is done in such manner as to insure good growth ("run") of mycelium through the compost.

During the next 15 to 30 days the grower needs only to maintain temperature, moisture and ventilation near optimum. Usually during this period the temperature will fall gradually to about 68°F. When the mycelium has grown and penetrated the compost sufficiently (3 to 4 inches from the points of inoculation), the bed is ready for casing. Casing is quite important because its purpose is to induce the production of the fruiting bodies of the mushroom. This process involves moistening the surface of the inoculated bed and then spreading a thin layer (0.5 to 1.5 inches) of soil on the surface. Not all soils are equally good as casing soils. Thus, Lambert[183] stated that heavy soils were better than sandy soils, and that clay loams were preferable when they were of such type that they did not cake on the bed. Heat-sterilized soils caused reduced yield, and Singer[266] has stated that steaming the casing soil can lower the yield.

Exactly why the addition of casing soil induces the initiation of basidiocarp formation by *Agaricus bisporus* has never been determined but it seems probable that since casing soil usually contains little organic matter, its effects are due to physical factors rather than to chemical factors. Gray[114] suggested that the effect of casing soil might be broadly explained by Klebs'[170] hypothesis that vegetation and reproduction are mutually antagonistic processes. On the basis of this hypothesis, the casing soil may provide an environment favorable for growth in all ways except that nutrients are lacking and that as a result the reproductive phase is initiated.

Some attempts have been made to find a substitute for natural soils for casing because (1) different soils vary in their effectiveness as casing soils, and (2) problems of soil deterioration are created by the removal of topsoil in areas where there is an extensive mushroom industry. Thus, Edwards and Flegg[90] have experimented with the use of mixtures of clay, vermiculite, sand and peat for casing purposes. A peat-vermiculite mixture (peat, 7.5 kg, vermiculite, 5.0 kg and lime, 2.0 kg) produced best mushroom yields when used with a moisture content of 42%.

Once the beds have been cased especial attention must be paid to ventilation, since as noted above, the accumulation of carbon dioxide in too great a concentration will inhibit or completely halt mushroom development. French growers have named the first sign that a bed is ready to produce mushrooms *la marque.* At this stage if the casing soil is slightly disturbed it will be seen to contain white rhizomorphs (white strands composed of bundles of mycelial filaments). It is on these rhizomorphs that the mushroom primordia (small compacted spheres of mycelia the size of a pinhead or smaller) develop. According to Singer[266] *la marque* will not be observed until at least 25 days after spawning, and delays of five months have been observed. However, he advises that if *la marque* has not appeared by the end of eight weeks, the bed should be checked for diseases, weed fungi, etc.

How soon a mushroom develops from a primordium depends to a very considerable extent upon the temperature at which the house is maintained and this can be turned to the advantage of the grower, since it may allow him to regulate the harvest time of his crop in such manner as to meet special market or labor conditions. Thus, Lambert[180] reported that basidiocarps of the white variety of *Agaricus bisporus* reach full growth in six days at 70°F but that at 50°F they reach full growth only after 22 days. Hence, the mushroom grower may enjoy a flexibility with respect to harvest time not enjoyed by most other agriculturalists, not at all a unique situation for the microbial farmer.

Unlike crops such as corn, wheat, potatoes, etc., mushrooms in a particular crop do not all appear at approximately the same time. On the contrary, several cycles ("breaks" or "flushes") of basidiocarp production occur at intervals. After the first break, which occurs over a limited period of time, a decline in production occurs and there will be a period of low production after which the second break begins to occur. This will be followed in like manner by several other breaks. Singer[266] reports that the total cropping period may last anywhere from one and one-half to seven months but that most growers stop after the fourth break. This is done to allow the grower to remove the spent compost and to clean the house in preparation for the next crop, since most American growers prefer to produce two crops per house per year. After the first several breaks, the breaks usually become successively smaller and further between, and hence attempts have been made to "rejuvenate" the beds by various means. One such method is the "flopover" in which the compost in the bed is turned over after the third or fourth break. Another method is to give the bed a "rest" by keeping it cool and dry. A third method is to remove the casing soil and a small amount of compost; the bed is then watered and in about five days is cased with fresh soil.

The time at which each mushroom is picked is quite important to the grower. If a mushroom is picked too early the grower loses because a young mushroom is still growing in weight and volume; if it is picked too late (after the pileus has fully expanded and the veil is broken thus exposing the gills) it will not reach the customer with the nice, fresh, white appearance of a younger picked mushroom. Singer[266] is of the opinion that "the culinary value of a fully mature mushroom is not equal to that attained before full maturity has been reached", but this reviewer does not concur in that opinion. If one compares the taste of the somewhat darkened mushrooms that have been allowed to reach maturity (but not pass it) so commonly encountered in England with the often quite innocuous tasting buttons of domestic origin, one must either conclude that truly mature mushrooms have a better taste or that somehow the English have succeeded in developing better cooks! Eye appeal is probably involved to a large extent, because the American housewife (many of whom never feel completely safe with fresh mushrooms anyhow) has been thoroughly educated to expect only young basidiocarps of the Snow White variety in the market. Another reason, however, for sending not quite mature mushrooms to market is that they have longer shelf life and are less susceptible to virus.

Since the construction of a mushroom house is expensive as is its ventilation, heating (and/or cooling) and maintenance, it is economically important to the grower to produce as many pounds of mushrooms per house as possible. Hence, since there is a maximum number of square feet of bed area that can be provided in a given house, the number of pounds of mushrooms which can be produced per square foot of bed area is a very important figure to the grower. In small scale experiments Lambert and Ayers[185] reported yields of slightly more than six pounds per square foot, but Singer[266] estimated that the average yield of mushrooms is one and one-half pounds per square foot of bed, one pound yields being rather poor and any yield above three pounds being very rare. These figures seem somewhat conservative since the average yield in the United States during the fiscal year 1967-68 was 2.18 pounds and for the fiscal year 1968-69 was 2.22 pounds (Table 1).

TABLE 1

Total United States Production of *Agaricus bisporus* in Fiscal 1967-68 and 1968-69

Fiscal year	Total production (lb)	Bed and tray area (sq ft)	Average yield/sq ft (lb)
1967-68	181×10^6	83×10^6	2.18
1968-69	189×10^6	85×10^6	2.22

Based on data of Bird.[25]

Reference has already been made to the fact that domestic mushroom production had an early beginning at Kennett Square, Pennsylvania, and this state still leads in domestic production. Of the 189,000,000 pounds of mushrooms produced in fiscal 1968-69, 121,000,000 pounds (64% of total domestic production) were produced in southeastern Pennsylvania. Bird[25] states that four-fifths of the 629 commercial growers in the United States are located in this area.

Rettew and Thompson[239] and Gray[114] have presented graphs which show that mushroom production rose in the United States from 10 million pounds in 1922 to about 62 million pounds in 1946. Figure 10 illustrates that since 1930 there has been a steady increase in domestic consumption of mushrooms. With greater prosperity and a steadily increasing population some increase in consumption would be expected; however, Figure 10 also shows that there has not only been an increase in total consumption but also a very marked increase in annual per capita consumption. In 1930 the average per capita consumption was 0.31 pounds, but that by fiscal 1967-68 this figure had risen to nearly 1.12 pounds (Bird[25]). From Figure 10 it will also be noted that total consumption in the United States was 225 million pounds in fiscal 1967-68. However, during this period domestic production was only 189 million pounds. Canned and dried mushrooms are now being imported in considerable quantities into the United States, the principal supplier being Taiwan, in which country the mushroom industry has developed entirely since 1960 (Houck[147]). Whether or not the American industry should attempt to expand to the point where it could at least meet domestic requirements is debatable, since Houck has noted that canning mushrooms can be grown in Taiwan for about 10 cents a pound, while in the United States the cost is about 27 cents. He estimated the total cost of canned mushrooms at 52 cents per pound in Taiwan as compared with a domestic cost of 70 cents per pound. Phenomenal growth of the Taiwan mushroom industry as reflected by the increased U.S. imports from this country is illustrated in Figure 11. Such increased imports are a matter of some concern in the

FIGURE 10

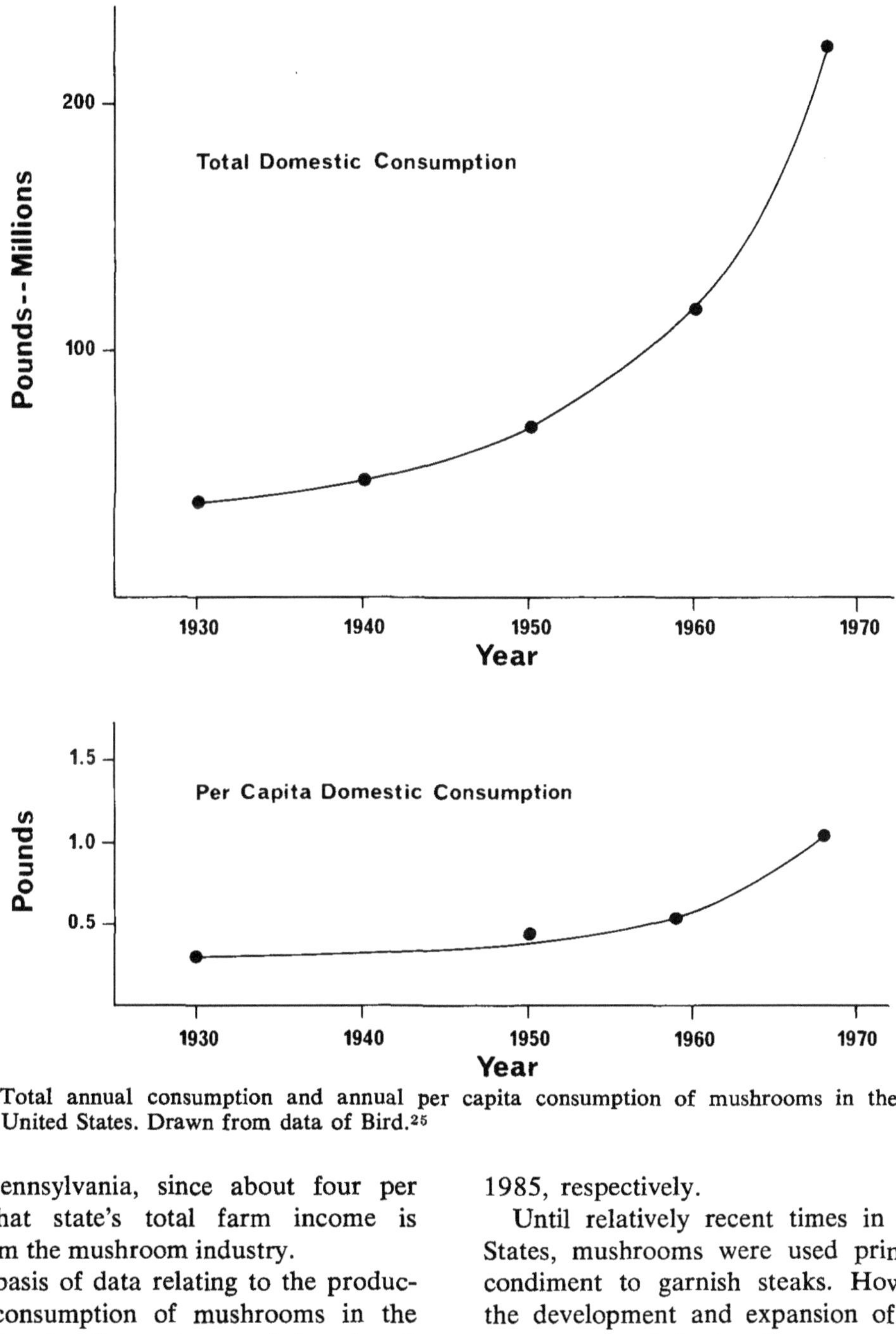

Total annual consumption and annual per capita consumption of mushrooms in the United States. Drawn from data of Bird.[25]

state of Pennsylvania, since about four per cent of that state's total farm income is derived from the mushroom industry.

On the basis of data relating to the production and consumption of mushrooms in the United States for the past several decades, it seems quite reasonable to predict that both production and consumption will continue to increase. Bird[25] projected his figures into the future and estimated that per capita consumption would be 1.57 pounds in 1976 and 2.07 pounds in 1985. In a later, less conservative estimate, he derived the figures 1.73 and 2.73 pounds as projections for the years 1976 and 1985, respectively.

Until relatively recent times in the United States, mushrooms were used primarily as a condiment to garnish steaks. However, with the development and expansion of the mushroom canning industry, they appear to be gaining favor as a base for soup and as an ingredient in many dishes in which they formerly were seldom used. For that reason it might be well to have a brief look at their food value. Skinner, Peterson and Steenbock[271] fed the cultivated mushroom to albino rats and found that they could digest 71% of the nitrogen in this mushroom, and Lintzel[190] reported that in

FIGURE 11

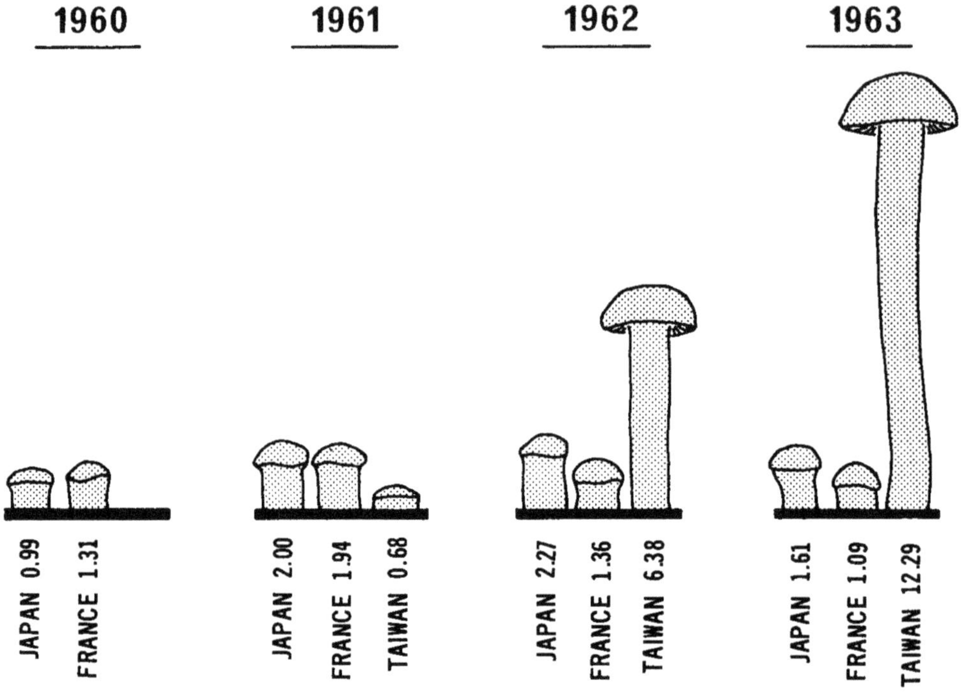

Development of the Taiwan mushroom industry as reflected in U.S. imports of mushrooms from that country. Redrawn from Houck, *Farm Economics,* February 1, 1964. Extension Service U.S.D.A. The Pennsylvania State University. With permission.

tests with humans the digestibility of mushrooms was 72-83%. Anderson[5] analyzed the Common Cultivated Mushroom and obtained the following results:

Water	98.50%
Protein (N × 6.25)	3.94%
Fat (ether extract)	0.19%
Extract matter	4.01%
Fiber	1.09%
Ash	1.26%
Calcium	0.0024%
Phosphorus	0.15%
Potassium	0.50%
Total iron	19.50 ppm
Available iron	5.95 ppm
Copper	1.35 ppm

Anderson and Fellers[6] reported that rats which received *Agaricus bisporus* as the sole source of protein survived a six week experimental period but made a weight gain of only 30 per cent of that attained by rats on a casein control diet. However, when they substituted mushroom protein for 20 per cent of the casein, a weight gain of 84 per cent of the control group was attained. They stated that mushroom protein is a partially complete protein such as wheat gliadin or barley hordein but that if properly supplemented, the mushroom is entirely suitable as a source of protein. These investigators also reported that mushrooms are an excellent source of nicotinic acid and riboflavin, a good source of pantothenic acid, and a fair source of vitamins B_1, C and K. McConnell and Esselen[208] also published a partial analysis of mushrooms and reported values close to those of Anderson and Fellers[6]: water, 88.9%; protein, 3.95%; fat, 0.26%; extract matter, 4.75%; fiber, 1.0%; and ash, 1.14%.

Analyses of one type or another have been performed by a variety of other investigators, but perhaps the best way to make a more practical assessment of the nutritive value of mushrooms is to compare their major constituents with those of a number of commonly-eaten vegetables as has been done in Table 2. In this table foods are listed in order of decreasing protein percentage, and if one or more foods contain the same percentage of protein, they are then arranged in order of decreasing number of Calories per 100 gram portion. From this listing of mushrooms and ten common vegetable foods, it is evident that based upon this arrangement, mushrooms occupy a position about in the middle. Thus, as sources of protein, mushrooms may be said to compare favorably with vegetables in general.

One further calculation might be illuminating. While total world consumption of mushrooms is not known, it is possible to calculate for the United States the contribution made by this type of food specifically in the area of supplying protein. Based on the assumption (Gray[115]) that on the average each individual should receive 65 grams of protein per day, the total annual protein requirements of the United States population (*ca.* 200 $\times$ 10^6) is 10,440,000,000 pounds. Using the more conservative value of Wooster[345] of 2.4 as the percentage of protein in mushrooms, the 225 million pounds of mushrooms consumed domestically in fiscal 1967-68 contained 5.4 million pounds of protein, a mere 0.052% of the total quantity of protein required by a population of 200 million.

According to Block et al.[28] mushroom propagation today has changed little from the process as described by Duggar[84] in 1904. This evaluation is a somewhat harsh one because since that early date many improvements in spawn making, strain selection, mushroom house construction, etc. have been made. However, as noted earlier, in spite of much experimentation on the development of a synthetic compost, the principal substrate ingredient for the cultivation of *Agaricus bisporus* is still composted horse manure as it was in Duggar's time. For that reason the works of Block[27] and Block et al.[29,31] are of especial interest. In their 1958 paper Block and his associates reported the results of experiments in which *Agaricus bisporus* was produced on compost, the principal ingredient of which was gum wood sawdust (Figure 12). The highest yield which they reported was 2.75 pounds per square foot, a figure somewhat higher than the national average; however, this yield was obtained during small-scale experimentation.

The results of the experiments reported by Block[27] in 1965 are even more suggestive. In these more recent experiments *Agaricus bisporus* was grown on composted fresh vegetable waste and newspaper (40:60) and cased with a mixture of equal volumes of sand and ground peat moss. The average yield from seven trays in two separate runs was 4.31 pounds per square foot, the highest single yield being 5.33. Block then tested composts made from various combinations of waste materials as well as commercially composted municipal garbage. Highest yield in these latter experiments was 5.11 pounds per square foot obtained on 40% vegetable waste and 60%

TABLE 2

Composition of Cultivated Mushrooms and Some Common Vegetables

Name	Calories/ 100 g	Water (%)	Protein (%)	Fat (%)	Carbohydrate (%)	Protein percentage (dry wt basis)
lima beans	128	66.5	7.5	0.8	23.5	22.5
green peas	98	74.3	6.7	0.4	17.7	26.1
chives	52	86.0	3.8	0.6	1.8	27.1
green beans	35	88.9	2.4	0.2	7.7	21.6
cauliflower	25	91.7	2.4	0.2	4.9	28.8
MUSHROOM	*16*	*91.1*	*2.4*	*0.3*	*4.0*	*26.9*
potatoes	83	73.8	2.0	0.1	19.1	7.6
beets (red)	42	87.6	1.6	0.1	9.6	12.9
cabbage	24	92.4	1.4	0.2	5.3	18.4
celery	18	93.7	1.3	0.2	3.7	20.6
egg plant	24	92.7	1.1	0.2	5.5	15.1

Based upon values listed by Wooster.[345]

FIGURE 12

The Common Cultivated Mushroom, *Agaricus bisporus*, growing on composted sawdust. Reprinted from Block et al., *J. Agr. Food Chem.*, 6, 923, 1958. Copyright 1958 by the American Chemical Society. Reprinted by permission of the copyright owner.

FIGURE 13

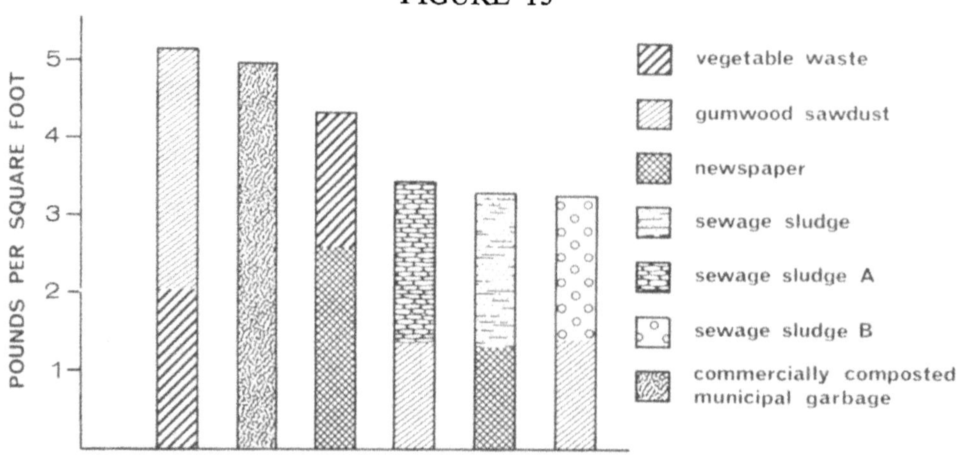

Pounds per square foot yields of *Agaricus bisporus* grown on various substrates. Based on data of Block.[27]

gum wood sawdust. A yield of 4.95 was obtained on one sample of commercially composted municipal garbage; but when similar composts from two other sources were used, the tests were unsuccessful. A summation of results on all substrates is presented diagrammatically in Figure 13.

Although the work of Block and his associates was conducted on a small scale, the results which they obtained were so promising that further research and development in the area of combining food production and waste disposal seem almost mandatory. A widespread use of garbage and newspaper in the manner suggested above would certainly be welcomed by most municipalities because of their ever-increasing waste disposal and pollution problems, and widespread use of sawdust would provide the timber industry with an outlet for a by-product for which there is no outlet in many areas today. In addition, a complete substitution for manure as the material of choice for composting would result in making available at least half a million tons of horse manure annually for use as fertilizer.

In the above brief account of the cultivation of *Agaricus bisporus* only the major points have been considered. Like any agriculturalist, the mushroom grower may be beset by many problems. While he does have an advantage over the corn or wheat farmer in that he can in large measure control the environment in which his crop develops, like these farmers his crop may become diseased, damaged by insect pests or crowded by "weed" fungi. For a more thorough discussion of this type of farming the reader should consult the works of Rettew and Thompson[239] and Singer[266] as well as the excellent review articles of Lambert[182] and Stoller[288] which have 166 and 219 literature citations, respectively.

Cultivation of the Padi Straw Mushroom

The Padi Straw Mushroom, widely grown for food in the Orient, appears in the literature under a variety of scientific names, but in the present discussion it will be referred to as *Volvariella volvacea.* Singer[266] has pointed out that to be more precise the varietal name *masseei* might be added and that a closely related cultivated form is *Volvariella diplasia.* He also states that there is a possibility that two other types (*Volvariella bresadolae* and *Volvariella volvacea* var. *heimii*) may also be cultivated. However, in the present discussion they will all be grouped together (not in the taxonomic sense but for ease of presentation) and referred to as *Volvariella volvacea.* The common name of this mushroom derives from the fact that it is most commonly cultured on rice (paddy) straw.

Like the Common Cultivated Mushroom, the Padi Straw Mushroom is an agaric in the older and broader sense (i.e., the hymenial layer is spread over lamellae or gills), but unlike *Agaricus bisporus* it is a member of the Family Amanitaceae rather than the Agaricaceae. Oddly enough, several poisonous mushrooms (Figure 14) are members of the Amanitaceae and one, *Amanita phalloides,* is considered to be the most deadly of all mushrooms. Like the poisonous Amanitas, the stipe of the Padi Straw Mushroom arises out of a basal cup (volva), and the fact that *Volvariella volvacea* has been an important article of the diet for many centuries provides ample evidence to refute the rather common myth that all mushrooms with a volva are poisonous.

Not only is Occidental literature concerning *Volvariella volvacea* quite scarce, but apparently Chinese literature on this subject is also not very abundant (Chang[49]). Therefore, in the present discussion, the reviewer will largely

FIGURE 14

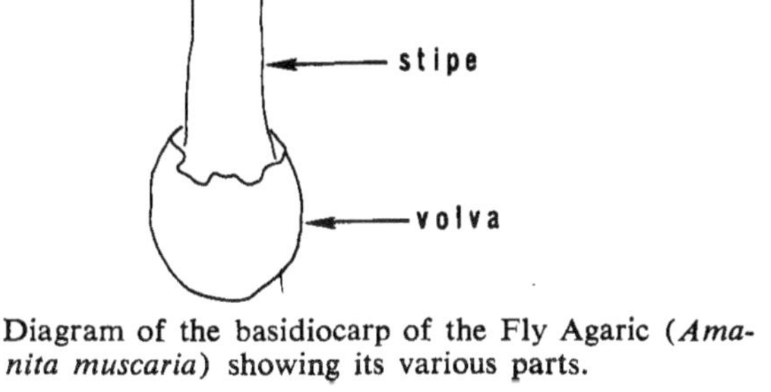

Diagram of the basidiocarp of the Fly Agaric (*Amanita muscaria*) showing its various parts.

depend upon the works of Singer[266] and those of Chang.[47-50]

In a very general sense the various steps in the cultivation of *Volvariella volvacea* are similar to those taken in the cultivation of *Agaricus bisporus*. However, the Padi Straw Mushroom industry has not been refined to any extent nearly approaching the degree of refinement which may be seen in the American mushroom industry. Although the history of *Volvariella volvacea* cultivation is not known, Singer[266] expresses the opinion that it is of very ancient origin, and that the rational growing of this species was originated by the Cantonese farmer.

Whereas in the United States virtually all mushroom growers use pure culture spawn, most growers of the Padi Straw Mushroom use spawn from "spent" beds to inoculate new straw beds. Apparently methods for the production of pure culture spawn have been developed (Su[291] and Chang[47]), but Chang[49] states that the peasants use "natural" spawn, viable mycelia found in spent straw from a previous cultivation. When the first crop of mushrooms is over, abandoned straw beds are cut into pieces and used to spawn other beds. At the end of the growing season, spawn from spent beds is stored for use in the following season by piling it up out of doors and covering it with straw or by keeping it indoors. Such methods of storage are obviously quite risky, and it seems probable that further refinement and expansion of the industry will depend upon more widespread use of pure culture spawn. Singer notes that just as in *Agaricus bisporus* cultivation, pure culture spawn of *Volvariella* is superior. Apparently pure culture spawn of *Volvariella volvacea* is prepared both by the tissue culture method of Duggar[85] and from spores as in the case of *Agaricus bisporus*. Chang[51,52] has recently initiated spore germination studies of *Volvariella volvacea* and, hence, it may be that in time it will be possible to select spawn with higher production capacities which will produce varieties of higher quality just as in the case of *Agaricus bisporus*.

Since much of the Padi Straw Mushroom production is accomplished by peasants on small holdings, each grower's practice may differ somewhat from others, although in the basic details the general procedures are the same. These consist of (1) land preparation, (2) soaking of straw, (3) construction of straw beds, (4) spawning, which is performed concurrently with bed construction, (5) care of beds, and (6) harvesting and marketing.

The land which is to be cropped is first plowed and then flooded for two or three days to exterminate worms and insects living in the soil. It is then drained and when dry enough to hoe is worked into elevated soil bases about one foot high. These soil bases are about 3 feet wide and 12 feet long and have paths about 1.5 feet wide between them. The paths are used by the grower when he tends the beds and when he harvests his crop, but in addition they may be flooded and thus serve as irrigation ditches. Chang[49] states that all beds are made in an east-west direction so as to provide more uniform conditions of sunlight and temperature. When the soil bases are constructed they are made higher in the center to permit drainage (Figure 15). Singer[266] gives the soil base dimensions as 3.5 x 16 feet and also states that the beds may be built on a thin layer of straw or on a cement floor where small-scale intensive culture is practiced.

The next step is the soaking of the straw upon which the mushrooms are to be grown, a step that is roughly analogous to the American and European practice of composting. Rice straw is the most common substrate for the culture of *Volvariella* (probably because it is the most common straw in many of the areas in which this mushroom is grown) although it will grow on other types of straw as well as

FIGURE 15

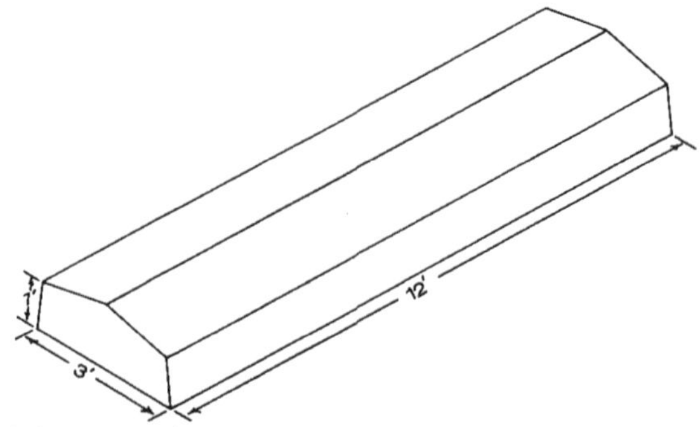

Diagrammatic representation of soil base on which straw beds are prepared for the cultivation of *Volvariella volvacea,* the Padi Straw Mushroom.

other materials. Sallet[249] states that *Bombax malabaricum* (cotton tree, kapok tree) is used as a substrate in Indo-China, and Bouriquet (*c.f.* Heim[139]) reported that clover distillation residues have been used as a substrate for *Volvariella volvacea* var. *heimii.* Prior to soaking, the straw is tied in bundles weighing about four pounds (Chang) or about three and one-half pounds (Singer). These two writers also differ in their descriptions of the "composting" process, ample evidence that different growers use different methods. According to Singer[266] the bundles of straw are immersed in water for 24 to 48 hours, being weighted down with bricks or rocks. On the contrary, Chang[49] states that the bundles are immersed in water until they are thoroughly soaked after which time they are removed and neatly placed in layers on the soil bases. As each layer is completed it is wetted with a mixture of about four gallons of distillery residues (stillage?) and ten pounds of rice bran to every 400 pounds of dry rice straw. Successive layers are placed on the pile until it is about 3.5 feet high, the straw being pressed down tightly. The temperature of the straw rises and is usually up to 64°C after three days. During this period the straw softens and turns brown and when yellowish, sticky fluid exudes from the pile, the straw is ready to be used in preparing the bed. Thus, the process of preparing the straw as described by Chang is more nearly comparable to the American composting process, while this step, as described by Singer, consists merely of soaking the straw.

The next step is that of constructing the straw beds. With the simple composting method described by Singer the straw bundles are next placed in layers on the soil base, and the straw bed is built to the desired height. Chang is not clear on this point, but it would appear that when the longer composting method is employed, the fermented straw is re-bundled and tied before being built into the straw bed.

In the cultivation of *Agaricus bisporus* the processes of preparation of bed and of spawning are two distinct and sequential processes. However, in the cultivation of *Volvariella volvacea,* bed construction and spawning are conducted simultaneously. Both Chang and Singer give specific directions for constructing each layer of straw bundles but these will not be repeated here because there are a great many ways in which the bundles could be placed. The objective is to place the first bundles of straw on the soil base in such manner as to form a fairly level layer with a thickness of one bundle of straw. Experience undoubtedly plays an important role here, because if the bundles are packed too loosely, the straw will dry out rather quickly, and if they are packed too tightly, aeration will be restricted and mycelial growth will be inhibited.

When the first layer has been prepared, pieces of spawn are placed about four to five inches apart, about five inches in from the edge of the layer. Singer[266] states that the third layer is not only spawned at the edges but all over the central area. Chang[49] recommends that each bed consist of five layers, but the number of layers per bed probably often varies with the individual grower. In some areas, where the cultivation of *Volvariella volvacea* is conducted in quite primitive fashion, the beds are not spawned at all but the grower depends upon chance infection of the straw by this species. At first sight, such primitive practice seems rather risky; however, it is quite possible that in an area where *Volvariella* cultivation has been practiced for a long period of time, this species might have become a rather prominent, perhaps even dominant member of the microbial population. Such a situation is not a unique one. For example, in the Shashi River Valley of Southern Rhodesia where the Sutu tribe has practiced palm wine making for many years, Boughey and Gray[36] found that a single yeast species (*Saccharomyces capensis*) was the predominant microorganism, not only in the fresh palm sap but also in the finished palm wine.

After the beds are constructed and spawned they are covered with straw mats or thatch and in some instances the upper part is covered with a waterproof cloth or plastic sheet during a rain. After a bed is constructed the temperature may rise to 50°C or higher, and water may be used judiciously to bring the temperature down gradually. Water must be added at intervals but again experience plays an important role. Singer[266] states that the beds should be kept "just moist" and that on the average small individual beds (3.5 x 3.5 feet) require about four and one-half gallons of water daily.

Fifteen to twenty-five days after beds are constructed, mushrooms may begin to appear (Figure 16), although the exact time required depends upon prevailing temperature and moisture conditions. In Chang's experiments the minimum air and bed temperatures were found to be 20° and 28°C, respectively. According to this investigator, best quality mushrooms were produced under conditions where the air temperature was 28°C, the relative humidity 85-95%, the bed temperature 34-37°C and the pH 5 to 6. Under these conditions about one week was required for a minute fruiting body to mature to harvestable size. The Padi Straw Mushroom fruits in rhythmic cycles much as does the Common Cultivated Mushroom. Therefore, after the first "break", several other successive breaks may occur at intervals of about ten days. Singer[266] states that the finest quality Padi Straw Mushrooms are those which are picked while the volva is still intact, and Chang[49] states that in Hong Kong it is customary to pick the mushrooms at this stage. The nutritional value of the Padi Straw Mushroom has not received quite as much attention as that of the Common Cultivated Mushroom; however, Chang[48] has published a partial analysis. These analyses, which are presented in Table 3, show that there is very little difference between mature mushrooms and those collected in the egg stage.

TABLE 3

Partial Analyses of Fresh Egg and Mature Stages of *Volvariella volvacea*

Percentages are averages of values obtained on fruiting bodies reared by three different methods.

Constituent	Egg stage	Mature stage
fiber	1.122	1.214
fat	0.529	0.582
protein	3.125	3.470
sugar	1.097	1.097

Based on data of Chang.[48]

FIGURE 16

The Padi Straw Mushroom, *Volvariella volvacea*, growing in a straw bed. Courtesy of Dr. S. T. Chang.

The writer has seen no estimates of the annual production of the Padi Straw Mushroom, but total production must reach quite sizeable proportions. Many are probably produced by the small grower for home use or for sale at a small local market and hence would never appear in any set of statistics, and in any event reliable estimates of Chinese production figures of any type have not been too abundant in the United States in recent years. This mushroom may be eaten fresh or may be dried for storage or for export, and Baker[16] has described the usual manner of drying as practiced in Malaya.

Expansion of the Padi Straw Mushroom industry should receive serious consideration in most of the grain growing areas of Asia. Since food in general and protein in particular have been in short supply on this continent for many years, increased production would lead to additional supplies of protein and in the process use materials (straw of various types) which presently have no value as human food. Several years ago the reviewer was informed of one program in Thailand, the express aim of which was to encourage the peasant to expand mushroom culture of this type; however, the fate of this program is presently not known.

Cultivation of the Shiitake in Eastern Asia

Like the Padi Straw Mushroom of the Orient and the Common Cultivated Mushroom of the Occident, the Shiitake has been referred to under a variety of different scientific names. Thus, at one time or another, it has been said to be a member of at least seven different genera of agarics. Among the names applied most commonly was *Cortinellus shiitake,* but as early as 1941 Singer[264] raised the question of whether Shiitake was a member of the genus *Cortinellus.* Singer[264,266] states that the correct name of the mushroom known under the common name of Shiitake is *Lentinus edodes,* and it is under this name that it will be referred to in the present discussion.

In the broad sense *Lentinus edodes* is an agaric, since the hymenial layer is spread over a lamellate (gilled) surface, but like *Volvariella volvacea* it is not a member of the family Agaricaceae. In older systems of classification both *Lentinus edodes* and *Volvariella volvacea,* as well as *Agaricus bisporus,* would be placed in the same family. However, in the more modern system of classification, *Lentinus edodes* is placed in the family Tricholomataceae, and, as was noted earlier, *Volvariella volvacea* is now placed in the family Amanitaceae, a taxonomically sound but psychologically unsound assignment because of the notoriously poisonous species which occur in this family.

In his 1941 description of *Lentinus edodes,* Singer[264] noted that this species grows on wood of dead deciduous trees mainly in the order Fagales (*Castanea*—chestnut, *Pasania*—shiia, *Quercus*—oak, *Carpinus*—hornbeam, *Fagus*—beech, but rarely on *Platycarva, Elaeocarpus* or *Magnolia*). Although the fungus obviously derived its common name because it was known to grow on shiia, it apparently shows no especial preference for this tree. *Lentinus edodes* is distributed naturally over eastern Asia and fruits all year around when temperature conditions permit. Since it is widely distributed through China, Japan and Indo-China, in all probability naturally-occurring Shiitake has been collected and eaten for many years, and Singer[266] notes that there are Japanese historical documents that refer to the eating of this mushroom as early as 199 A.D. Apparently, then, some form of cultivation of *Lentinus edodes* has been practiced for many years, but the work of Kondō and Kasahara[176] and Nisikado and Yamauti[220] on spore germination and sexuality in this species has helped place its cultivation on a somewhat more scientific basis. Such studies lead to methods for better spawn production and as may be seen with the Common Cultivated Mushroom, any developments which lead to improvements in spawn making usually lead to improvements in the industry.

The fact that Shiitake grows on wood would indicate that with a substrate so different from composted manure or soaked straw, the details of its culture would differ from those of either the Common Cultivated Mushroom or the Padi Straw Mushroom. In a very general way, however, the steps taken in connection with cultivation of Shiitake are analogous with those taken in connection with cultivation of the above two types except that there is no step that would compare closely with manure composting or the soaking of bundles of straw. The following discussion of its cultivation is largely based upon the observations of Singer.[266]

As noted earlier, in the more primitive forms of cultivation of the Padi Straw Mushroom, no spawn is used. The substrate (bundles of straw built into beds) is prepared and the inoculation is left entirely to chance infection by the desired fungus. Such a situation also obtained in the case of the earlier and more primitive cultivation of Shiitake, although inoculation was not left entirely to chance since logs that were to be inoculated were placed close to logs already permeated with mycelium of *Lentinus edodes.* In modern Shiitake growing, more frequently than not, inoculation is performed using a spore emulsion applied directly to the logs and thus spawn production as such is not involved. However, spawn is sometimes used, the mycelium being cultured on a medium consisting of sawdust from the proper species of tree plus two per cent rice bran. Spawn is also prepared by culturing mycelium on autoclaved, small, wedge-shaped or cylindrical pieces of oak wood.

Whereas in the cultivation of *Agaricus bisporus* or *Volvariella volvacea* the grower prepares beds of composted horse manure or soaked bundles of straw, the "beds" on which *Lentinus edodes* are grown consist of small logs placed in a suitable fashion. Singer[266] lists five species of *Quercus,* two of *Pasania,* one of *Castanea,* four of *Cyclobalanopsis* and two of *Carpinus* which are suitable for the preparation of logs for the growth of Shiitake but notes that three species of *Alnus* and one of maple (*Acer pictum*) are sometimes used, but more rarely. He briefly discusses the relative merits of these various species as substrates but recommends *Quercus serrata or Quercus acutissima* as the source of the two choice woods. Probably the choice of wood used by the small grower is determined in large measure by what is available to him in his area. It is obvious that in an area such as east Asia it would not be economical or feasible to use logs of such dimensions that they could be used for lumber purposes, an observation borne out by the dimensions listed by Singer (3-5 feet long x 2-6 inches in diameter). Thus, it would appear that logs from small, shrubby trees or the tops and larger branches of trees cut for lumber purposes form the principal source of substrate for the Shiitake grower. The trees to be used are cut and left in the woods until time for inoculation, at which time they are cut into logs of proper size.

Method of inoculation depends upon type of inoculum used. In the more primitive method, newly-cut logs are simply placed next to logs already invaded with mycelium of *Lentinus edodes*. If spawn is used, holes are drilled or pounded into the wood and pieces of spawn (wedge-shaped or cylindrical pieces of wood permeated with mycelium or small amounts of sawdust-bran spawn) are inserted. That the industry is one of long standing is evidenced by the fact that special hammers for making wedge-shaped or cylindrical holes in the logs have been developed.

The nearest approach to the step of composting, which is routinely practiced in the Occidental mushroom industry, occurs in the Shiitake growing industry when the grower uses a spore emulsion as inoculum. If such is the case, the logs are prepared by first soaking in water. The bark is then pounded and the spore emulsion is applied by pouring, injecting or spraying over the incisions made into the sapwood.

After being inoculated the logs are placed in the "laying yard," a site judged to be favorable for the development of the mycelium in the wood. There the logs are laid at a small angle to the surface of the soil (Figure 17). Selection of a site for the laying yard is important, since it is during the laying period that mycelium develops well or poorly in the logs, and size, quality, number, and time of appearance of the subsequent crop of mushrooms will depend upon the production of logs well-permeated with mycelium. What occurs in the laying yard is, of course, comparable to the run of mycelium of *Agaricus bisporus* through a bed of composted manure. In an unsuitable site, mycelium development may be poor or the logs may be invaded and permeated with mycelia of "weed" fungi (i.e., undesirable or inedible wood-inhabiting species). In a properly selected site the logs may require little or no watering except during a dry season. Usually the laying operation (i.e., development of mycelia through the logs) requires five to eight months, although in an outdoor operation of this type the length of time required will depend upon prevailing environmental conditions.

FIGURE 17

Method of placing logs in the "laying yard" for the cultivation of the Shiitake, *Lentinus edodes*.

At the end of the laying period the now-permeated logs are transferred to the "raising yard". Singer likens this transfer to the casing operation in the Common Cultivated Mushroom industry, and so it is in the sense that the fungus is now placed in a situation more favorable for reproduction than for growth. Moisture requirements for basidiocarp production of *Lentinus edodes* are higher for reproduction than for growth and hence the raising yard must be better shaded to restrict moisture loss, and watering is required more frequently than during the laying period. Singer[266] recommends the placing of logs in the raising yard during the winter so that they will be ready for cropping in early spring, since the desirable temperature for fruiting is 12-20°C. In the raising yard the logs are placed in a more

FIGURE 18

Method of placing logs in the "raising yard" for the cultivation of the Shiitake, *Lentinus edodes*.

upright position than they were in the laying yard, generally being placed against rows of bamboo fences (Figure 18), although in smaller and more primitive operations they may simply be leaned against standing trees in much the same fashion as the midwestern farmer stacks freshly-cut fence posts to season. Apparently some Shiitake is grown indoors, since Singer has reproduced two photographs in which Shiitake is shown growing on logs in a greenhouse.

Both spring and fall crops are produced on the same logs, but the spring crop is often more luxuriant although more viable spores are obtained from the fall crop (Singer[266]). The Shiitake grower enjoys a much longer "bed" life than does either the grower of *Agaricus bisporus* or *Volvariella volvacea*. Logs produce for a minimum of three years and may continue in production for as long as six years. Thus, with two crops per year, the grower harvests from six to twelve crops from the same logs before he has to replace them with freshly-inoculated logs. Logs which are producing must be kept moist but otherwise they need no attention except for daily harvesting. Ultimately all of the nutritive materials in a log are used up by the mycelium and at that time the log may be said to be "spent" in the same sense that compost or bundles of straw are referred to as spent.

Harvesting and packing of Shiitake do not require the care required by *Agaricus bisporus*. Growers of the latter species prefer to harvest a basidiocarp before the veil is broken and the pileus fully expanded, because the spores of this species become dark when they mature and thus in a fully-developed mushroom the lamellae on which the spores are borne are dark in color. This problem is not encountered with *Lentinus edodes*, since the spores do not become dark at maturity. Hence, even in fully expanded basidiocarps, the lamellae are not dark in color. The grower of *Lentinus edodes* enjoys still another advantage not available to the grower of *Agaricus bisporus*. This latter species may be bruised or scratched very easily and when this occurs the basidiocarp may become discolored due to melanins produced through the activity of polyphenol oxidases present in some abundance in *Agaricus bisporus*. The context of *Lentinus edodes* is much more firm and hence is not damaged so easily. In addition, it apparently does not contain oxidases of the type found in *Agaricus bisporus* so discoloration of mushrooms due to slight injuries does not occur.

Most Shiitake is sold in the dried condition (Figure 19), and it may be dried using artificial heat or may be sun-dried. The sizeable proportions of the industry are exemplified by the fact that in 1953 over three and one-fourth million pounds of this mushroom were exported from Japan to Taiwan, Hong Kong, Malaya, Singapore, the United States and other countries (Singer[266]). The extent of current production is not known to the writer, but since Japan has had a long history in the use of fungi directly as food or in food processing, it seems reasonable to assume that the Shiitake industry is a large one. On the other hand, with the increasing industrialization of Japan it may well be that the labor required in producing Shiitake might result in the acquiring of greater amounts of foreign currency if it was channeled into other industries such as shipbuilding, electronics, and the many other industries which the Japanese have developed so extensively since World War II.

To anyone who has witnessed typical lumbering operations in the eastern deciduous forests of the United States, the above brief discussion of Shiitake cultivation probably suggests that a similar operation could be conducted in this country using as substrate the slash so commonly left from a cutting operation. Whether or not *Lentinus edodes* could be grown successfully in the United States is not

FIGURE 19

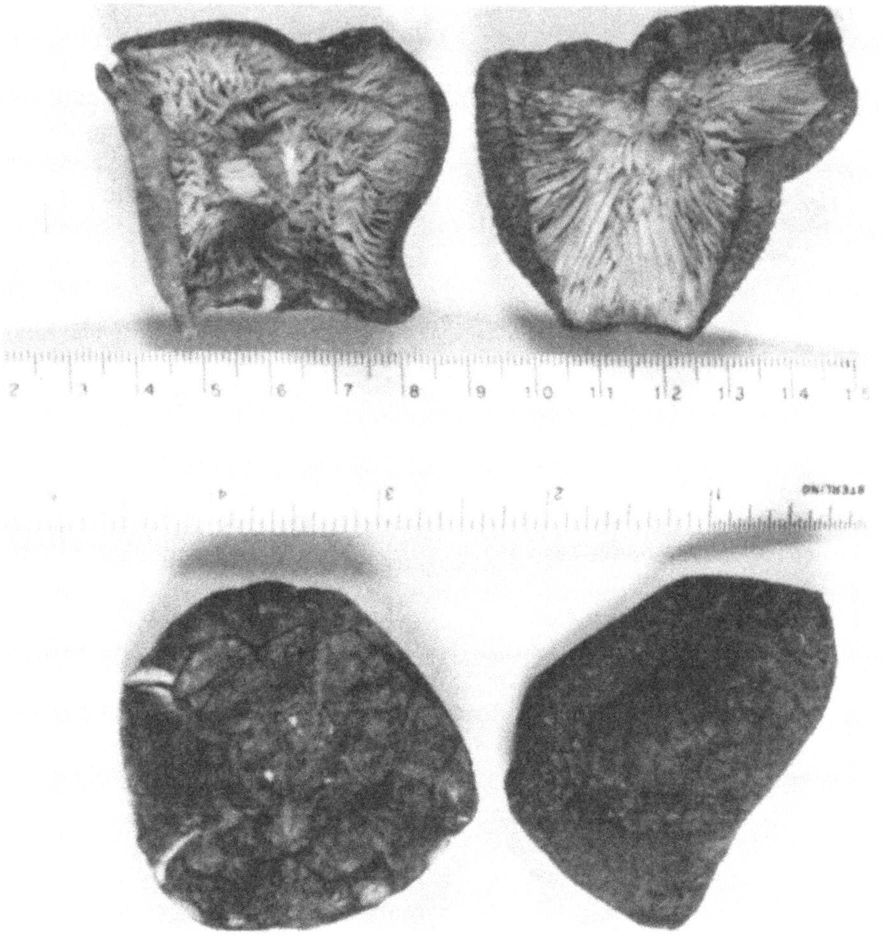

Dried Shiitake (*Lentinus edodes*) as marketed, showing top and bottom views of pilei.

known, but Christensen[58] lists two edible domestic species of *Lentinus* and there are other edible wood-inhabiting mushrooms which might conceivably be used. Certainly Passecker[227] has had *Lentinus edodes* in pure culture in Vienna, and in this writer's mind there is little question but that there are areas in the United States where this or related species could be grown. The greatest present obstacle to the large-scale cultivation of an edible wood-inhabiting mushroom in the United States is the high cost of such an operation. What with the great amount of hand labor involved and the present high cost of labor, such an operation would be so costly that any mushrooms so produced would be priced right out of the market. Looking to the future, however, when (1) the continued addition of about three and one-half million new citizens annually leads to real population pressures, (2) the continued irresponsible dissipation of our natural resources leads to many shortages, (3) the increased pollution of our environment with all sorts of solid and gaseous wastes makes waste disposal an even greater problem, and (4) insufficient amounts and unequal distribution of protein make malnutrition even more widespread than it is currently, we may then see wood, now left to rot on the ground, being used to produce part of the food that an increased population will require. In the following section some experimental work is described that suggests the possibility of using wood and wood products for the production of mushrooms.

FIGURE 20

Chinese Ear Fungus—*Auricularia polytricha,* showing cluster of moist, swollen basidiocarps (above) and dry basidiocarps (below).

Potential Cultivation of Other Mushrooms

Mycologists have long known that there are many edible types of fungi other than the four just discussed in connection with their cultivation in various parts of the world. Hence, over the years there have been many suggestions that one species or another should be brought into cultivation; in most such cases it has simply remained a suggestion. In addition to the four species already mentioned, several have been cultivated on a small or primitive scale or experimentally. For example, one of the Ear Fungi, *Auricularia polytricha,* has been used as food in China for many years. Like *Lentinus edodes, Auricularia polytricha* as well as other related species, grows on wood and, when fresh, is thick, gelatinous, brown in color and often roughly shaped like an ear. Also like *Lentinus, Agaricus* and *Volvariella, Auricularia* is a member of the Basidiomycetes, but it is not an agaric even in the broad sense. The Ear Fungus dries to a brittle, paper-thin sheet but may be reconstituted by soaking in water (Figure 20). It has long been possible to obtain imported dried *Auricularia* in San Francisco and other places on the west coast, and it is sometimes seen offered for sale in other parts of the United States. However, it is usually not considered a particularly desirable food item except by the Chinese. Ear Fungi are collected for use from their wild natural habitats but apparently in some instances a primitive cultivation on wood is conducted. At this time it seems highly unlikely that this particular type of fungus would ever become very popular in the United States. However, since it is highly prized in China it seems probable that efforts might be made to place its cultivation on a more scientific basis.

Singer[266] states that cultivation of the Fairy Ring Mushroom (*Marasmius oreades*) has been practiced by individual gardeners in Canada and England for over 50 years, but the only description of the cultivation of this species is the early report of Buller.[39] The details of Buller's method will not be repeated here since it obviously did not produce especially promising results. In a laboriously prepared outdoor bed with 174 square feet of surface, the first crop appeared in the second year and during this and the succeeding three years a total of 20 baskets (14 x 7.5 x 5 inches in size) of mushrooms was obtained. Considering the effort required to prepare the bed, weed it and spade it, five years seems to be an inordinately long period of time to wait for such a small harvest. In a normal *Agaricus bisporus* operation, 174 square feet of bed space should yield at least 350 pounds of mushrooms in less than six months.

Because of their delicious flavor, attempts have been made to cultivate various species of morels. Morels appear in the early spring (in 34 years of collecting in Ohio, Dr. W. G. Stover found morels in the autumn only once), so the temperatures at which morel fruiting bodies (ascocarps) will be formed are obviously rather low. Among those laymen who regularly hunt morels there exists a considerable difference of opinion concerning the best habitats for these mushrooms. Thus, one collector will state that they are more likely to be found near old apple trees, while another will claim that they occur primarily under ash trees (the writer has found them in abundance in both habitats). Singer[266] states the *Morchella esculenta* is found most frequently near elms. However, in central Ohio the present writer found this and several other species in greatest abundance in beech-maple woods. It would probably be safest to say that morels are where you find them, since several years ago the writer received a photograph in which morels were shown forming their fruiting bodies on exposed soil in a basement floor where some of the concrete flooring had broken away. It has been suggested that morels may form mycorrhizae with the roots of trees and that in some instances they may parasitize plants, but these claims need further substantiation. Brock[38] studied the nutrition of *Morchella esculenta* but similar studies on other species are needed. Molliard (*c.f.* Singer[266]) grew morels (*Morchella esculenta* and *Morchella hortensis*) out of doors under cherry trees in beds prepared from apple residue and old papers as well as in pots in the greenhouse. Babee (*c.f.* Singer[266]) reported that he had cultivated *Morchella costata* near woods on garbage mixed with garden soil, and Heim[138] and Costantin[65] repeated Molliard's earlier work with *Morchella hortensis*. Costantin reported a yield of 350 per square meter of bed surface which calculates to almost 0.07 pounds per square foot. Delicious as the morel may be and desirable as its large-scale cultivation may seem, its economical production must await the development of cultural practices that will result in yields greater than one pound per fourteen square feet. Bode[33] has suggested a unique culture method for the production of morels. Since morels grow well on the tubers of Jerusalem artichoke (which have high inulin content) this investigator recommended the introduction of morels into artichoke fields; thus, if no morels form, there is always a crop of artichokes which can be harvested.

Various other fungi have been cultivated on a small scale or have been proposed as potentially valuable species for cultivation. However, most of these need not be discussed here, since there seems to be little justification for attempting to cultivate additional species unless (1) they are as well-flavored or are better-flavored than presently produced mushrooms, (2) they can be produced economically in large quantities, and (3) if possible various undesirable wastes can be disposed of or rendered innocuous in the course of their production. Because of the third point above, the work of Block et al.[29–31] seems especially promising.

Liese[189] successfully cultivated the Oyster Mushroom (*Pleurotus ostreatus*) on beech wood in much the same way as Shiitake is cultivated, but Singer[266] expressed doubts that further attempts should be made to bring this species into commercial production, presumably because it is a mild parasite of fruit and forest trees. The writer has never seen *Pleurotus ostreatus* growing on living trees except those which were in a weakened condition or already severely damaged; however, it would be prudent not to introduce this species into a forest where it does not already occur. Block[29–31] and his associates have proposed a method for growing this species, however, which would not involve the introduction of a potential parasite into a stand of living trees as would have to be done if it was cultivated like Shiitake. These workers first prepared eleven different media, the principal ingredients of which were pine, oak or gum wood sawdust, although in one instance bagasse was used. Portions of each type of medium were inoculated with each of the following wood-destroying Basidiomycetes: *Pleurotus ostreatus, Collybia velutipes, Lentinus lepideus, Agaricus blazei, Polyporus sulphureus, Clitocybe tabescens, Polyporus frondosus, Hydnum coralloides* and *Armillaria mellea.* Of these nine species, *Pleurotus ostreatus* grew more vigorously on most of the media and produced fruiting bodies more readily than did the others. This species was then cultured on pine and gum wood sawdusts (Figure 21) fortified

FIGURE 21

The Oyster Mushroom (*Pleurotus ostreatus*) growing on balsa wood sawdust. From Block et al. *J. Agr. Food Chem.*, 6, 923, 1958. Copyright 1958 by The American Chemical Society. Reprinted by permission of the copyright owner.

with five per cent oatmeal. Yields on gum wood sawdust were about twice as large as those obtained on pine sawdust, which might have been expected since the usual habitat of this species is wood of deciduous trees. When cultured on fortified gum wood sawdust, the average yield of fresh mushrooms was 0.48 g fresh mushrooms per gram dry ingredients. In a final series of experiments on a balsa wood compost containing soybean meal and mineral salts, a yield of 1.41 g of fresh mushrooms per gram dry ingredients was obtained. In their discussion of a short method of composting, Sinden and Hauser[263] cite one experiment in which a yield of 836 pounds of mushrooms (*Agaricus bisporus*) was obtained per ton of manure. On the assumption that manure contained 50% water, this yield was about 0.84 pound per pound of dry matter. Thus, it would appear that Block's yield of 1.41 pounds of *Pleurotus ostreatus* was much superior to Sinden and Hauser's 0.84 pound of *Agaricus bisporus* per pound of dry material. However, Sinden and Hauser obtained 2.16 pounds per square foot; whereas under the experimental conditions of Block and his associates, yield calculated in this manner was only 0.78 pounds per square foot. Thus, to obtain the same volume of production (equal weights of fresh mushrooms), three times as much bed area would be required for producing *Pleurotus ostreatus* on sawdust as for producing *Agaricus bisporus* on composted manure, a requirement which probably would render production of the first species uneconomical. Nonetheless there is no reason to believe that further experimentation would not lead to increased yields. These workers state that the Oyster Mushroom does not have the thick "meaty" tissue of buttons of the Common Cultivated Mushroom. It is quite true that the textures of these two mushrooms are quite different; however, in the writer's opinion the flavor of the Oyster Mushroom is superior.

Fungi as Animal Food

Most mushroom collectors have found frequent evidence in the form of carpophores that have been partially eaten that animals of one type or another occasionally use various fleshy fungi as food. However, there are few documented instances of fungi being deliberately fed to animals (other than experimentally) or of fungi constituting any sizeable part of the animal diet. In his discussion of the fungus flora of the Arctic, Singer[265] records that in arctic and subarctic areas reindeer feed on mushrooms, even digging frozen ones from under the snow, and that the local and seasonal abundance of mushrooms is one factor in choosing feeding grounds in Lapland and in the northern U.S.S.R. From Finland, Rautavaara[237] reports that in a good mushroom year reindeer weigh on the average 20% higher than normal while in a year when mushrooms are scarce they weigh 12% less. According to this investigator about 20% of the meat production can be attributed to the feeding of fodder containing mushrooms, and that in Finnish Lapland alone, this meat is valued at about ten million Finnish marks annually.

Singer[266] states that in Germany, during and after World War I, wild mushrooms were successfully fed to pigs, poultry and fish. This same worker suggests that the waste material from mushroom canneries (lower portion of stipes, damaged or diseased mushrooms) could be used as fodder but cites no instance of this having been done. The amount of such waste material is relatively small and could make only a negligible contribution to the supply of animal feed. Also, the expense involved in the production of mushrooms is too great to permit their use as animal feed, so at this time the only areas where fungi play a significant role in the feeding of animals are in the far north.

USE OF FUNGI IN FOOD PROCESSING

The previous section was concerned with those fungi (and methods for culturing them) which are consumed directly as food. However, fungi have still other uses in the food industry, because in addition to these direct uses, fungi are also used to process certain materials in the production of various specific foodstuffs. Needless to say, different types of fungi are used for food processing than those which are used directly as food. In the latter instance only those fungi which produce a fleshy, relatively large reproductive structure are used, while in the processing of food, mold types of fungi are commonly employed. When fungi are used in food processing the processor is not primarily interested in the presence of the fungus *per se* in the resultant product. On the contrary, he is interested in the changes which the fungus has effected in the substrate. Thus, foods of desired texture, flavor, odor, digestibility and keeping quality are produced and, while fungus mycelium and spores may be in the finished product, their actual presence is of secondary importance in a consideration of these foods.

Fungi in the Processing of Cheese

Who first discovered a process for making cheese and exactly when the discovery was made is not known, but this discovery unquestionably constituted a major step forward in food technology as well as in human nutrition. Prior to the discovery of a cheese-making process there was no known method for preserving the important nutrients of milk for any considerable period of time and, until the advent of modern methods of food handling and food processing, cheese constituted the only means of storing milk proteins. Scott[255] states that Pliny praised the mold-ripened cheese, Roquefort, in the first century A.D., so it is probable that cheese of some type was prepared many years before Pliny's time since the odds are very great against a mold-ripened cheese having been prepared prior to bacterially-ripened cheeses.

Since the initial discovery (or more probably many independent discoveries) of a process(es) for making cheese, man has shown considerable initiative in this area of food processing as is illustrated by the fact that by 1918 Doane and Lawson[79] could list nearly three hundred different varieties. This number increased rapidly and in 1925 Thom and Fisk[306] were able to list over five hundred varieties. Although there are a great many named varieties of cheese, they may all be classified into a relatively small number of general categories. Such a classification is presented in Thom and Fisk[306] and Gray.[114] Unfortunately much of the cheese making effort in the United States in recent years seems to have been directed toward the production of a greater number of bland, innocuous, often rather unpleasantly textured materials known as process cheeses. Although it is still possible to obtain a rather wide variety of distinctive and delicious true cheeses in England and on the continent of Europe, in England at least, more and more process cheeses seem to be making their appearance.

As one examines a classification of the various types of cheese it soon becomes apparent that there is a much greater number of bacterially-ripened cheeses than there is of fungus ripened cheeses. Although there is a considerable number of named cheese varieties, all of which are ripened by fungi, these can actually be grouped into two basic types: the Roquefort (blue or blue-veined) type and the Camembert type. Thus, in France the best known and probably oldest known blue cheese is Roquefort, but other cheeses such as Pâté Bleu, Septmoncel, Gex, Mont Cenis, etc. are all of the Roquefort type. Similarly the Gorgonzola cheese of Italy and the Stilton cheese of England are also of the Roquefort type. Scott[255] recognizes five named varieties of blue-veined cheeses from the United Kingdom, five from Italy, fourteen from France, and one each from Switzerland, Spain, Portugal and Greece.

Only the two basic types of fungus-processed cheese will be considered in the present discussion, since many varieties have been named merely on the basis of the locality in which they were first made. A comparison of (1) the methods of processing Roquefort and Camembert cheeses, (2) the species of fungi involved, (3) the manner in which the fungus grows in relation to the curd, and (4) the changes brought about by the fungus during ripening well-illustrate that there are fundamental differences between these two fungus-ripened cheeses that warrant their being considered distinct basic types.

Roquefort Cheese

Roquefort-type cheese originated in southern France many years ago and in its original form was made only from sheep's milk. Today most Roquefort and all of its imitations are made from cow's milk. Thom and Currie[305] early reported that a specific fungus, *Penicillium roqueforti* occurs in practically pure culture in Roquefort, Gorgonzola and Stilton cheeses. This was verified by Dattilo-Rubbo[72] who isolated fungi from eight blue-veined cheeses (Stilton, Blue Cheshire, Roquefort, Wensleydale, Danish Roquefort, Gorgonzola, Blue Vinney and Dolce Verde) and found *Penicillium roqueforti* in all of them except Dolce Verde. The Dolce Verde fungus was described as being related to *Penicillium expansum,* a fungus associated with fruit decay. Although Dolce Verde may represent an exceptional case, in general it may be stated that *Penicillium roqueforti* is responsible for the ripening of blue-veined cheeses.

In making Roquefort cheese, the curd is inoculated with *Penicillium roqueforti* by mixing in bread crumbs on which the organism has been growing. Thom and Currie[305] stated that Gorgonzola and Stilton cheeses are not inoculated, but Doane and Lawson[79] claimed that in the manufacture of Gorgonzola, layers of curd are interspersed with moldy bread crumbs. In all probability most blue-veined cheeses are inoculated with the proper fungus, although in a factory where such cheeses have been made for any considerable period of time, in all likelihood the dominant member of the microbial population is the desired fungus and hence inoculation of the curd might not be necessary in all instances. Thus, Scott[255] states that spores of varieties of *Penicillium* predominate in the atmosphere of cow sheds, farm dairies and cheese rooms. However, he notes that since very clean milk may lack these spores or when a dairy has ceased cheese production for some years (i.e., during wartime), cheese may fail to "blue." He then states that it is common practice (in England) to add mold cultures to milk or to dust the curd with spores. Since the invention of Stilton cheese in England has been attributed to a Mrs. Paulet or a Mrs. Orton in the seventeenth century, it is evident that by the time that Thom and Currie[305] studied the manufacture of Stilton cheese in 1913, its production was a matter of long years standing and inoculation of the curd in older factories was probably unnecessary. However, cheese making ceased in some dairies for several years during World War II and some time may have elapsed before the desired mold variety was once again established in a factory. Leaving inoculation of the curd to chance may be somewhat risky, since Scott notes that sometimes "white" Stilton may be a defective cheese which failed to blue.

After inoculation the unripened cheeses are placed in ripening rooms where temperature and humidity are carefully controlled and there left until the fungus has grown to the desired extent. They are then placed at lower temperature where growth is inhibited but fungus enzymes remain active. Since filamentous fungi in general are strongly aerobic organisms, the question naturally arises as to how *Penicillium roqueforti* is able to grow in the interior of the pressed curd. Thom and Currie[305] demonstrated the unique capabilities of this species many years ago by showing that it could grow at high carbon dioxide concentrations and low oxygen concentrations, conditions under which many other species could not grow. Thus, in an atmosphere consisting of 75% carbon dioxide and 25% air, of 22 species of *Aspergillus* and *Penicillium* tested, only *Penicillium roqueforti* formed fairly strong colonies. That the oxygen requirements of *Penicillium roqueforti* are quite low is evidenced by the findings of Golding[111-113] that this species would grow in an atmosphere with oxygen concentration as low as 4.25%. Nonetheless, this species needs some oxygen and for that reason holes are punched in the pressed curds to provide means of aeration deep in the curd.

Anyone familiar with blue-veined cheeses will agree that they have a peculiarly unique flavor and aroma. These characteristics are undoubtedly due to the action of the fungus ripening agent, and the nature of the responsible material(s) has concerned a number of investigators. Jensen[160] suggested that the flavor and aroma were due to ethyl butyrate but since he presented no analytical data to substantiate this view it must be relegated to the realm of speculation. Currie[67] reported that during the ripening of Roquefort cheese, *Penicillium roqueforti* produces a water soluble lipase which catalyzes the hydrolysis of milk fats. He recovered acetic, butyric, capric, caprylic and caproic acids by steam distillation. Currie stated that of these acids capric, caprylic and caproic and their readily hydrolyzable salts have a "peppery" taste and are responsible for the burning effect of Roquefort cheese on the tongue and palate. Ironically, and probably to the distress of the gourmet when he learns it, these same acids are found in goat perspiration which obviously bears little resemblance to a good blue-veined cheese. Scott[255] states that in some of the newer varieties of blue cheese, cultures of *Candida lipolytica* are added to provide additional lipolytic activity. However, he advises caution in selection of the strain of *Candida,* since a too active strain may inhibit mold growth or produce rancid flavors.

Hammer and Bryant[133] believed that the taste and odor of blue cheese are suggestive of 2-heptanone and that this methyl ketone apparently is formed from caprylic acid through the action of *Penicillium roqueforti.* Using an ether extraction procedure Patton[229] recovered material from blue cheese which contained a high concentration of methyl ketones. Fractional distillation yielded relatively pure fractions of 2-pentanone, 2-heptanone and 2-nonanone, and it was suggested that these methyl ketones are formed from the fatty acids in blue cheese by beta-oxidation. Patton pointed out that a number of observers had noted the similarity in odor between these ketones (and particularly 2-heptanone) and blue cheeses and suggested that the unique flavor of these cheeses was probably due to minute quantities of methyl ketones. Girolami and Knight[109] studied fatty acid oxidation by *Penicillium roqueforti.* Preformed mycelia were capable of oxidizing fatty acids to methyl ketones of one less carbon atom than the acid from which they were formed. These ketones, which are listed in Table 4, were not further metabolized by *Penicillium roqueforti* in their experiments.

In view of such findings as those of Hammer and Bryant,[133] Patton[229] and Girolami and Knight,[109] it would appear that certain methyl ketones, and especially 2-heptanone, play a major role in imparting the characteris-

TABLE 4

Production of Methyl Ketones by *Penicillium roqueforti*

Substrate	Ketone
butyrate	acetone
valerate	2-butanone
caproate	2-pentanone
heptylate	2-hexanone
caprylate	2-heptanone
pelargonate	2-octanone(?)

From Girolami and Knight.[109]

tic odor and flavor to blue-veined cheeses. However, the fatty acids reported by Currie[67] may also contribute, and in all probability the subtle differences often noticeable between two cheeses made at the same time in the same factory may be due to minute differences in the concentrations and proportions of all these materials.

That changes be recommended in procedures for making a centuries old food product is inevitable. Thibodeau and Macy (*c.f.* Scott[255]) studied the lipase and protease systems of *Penicillium roqueforti* and were able to extract enzymes from the mycelium. The addition of such enzymes in powder form to the curd shortened the time required for ripening by several months. While gourmets may regard process blue cheese with distaste. the fact still remains that more and more process cheese is appearing in U.S. markets and blue cheese is no exception. Process blue cheeses (spreads, etc.) are generally lighter in color than traditional blue cheeses and this may be due to the use of lighter-colored or white variants of *Penicillium roqueforti.* Knight et al.[172] obtained a white strain of this species by ultraviolet irradiation, and Morris et al.[214] found that cheese ripened with white strains developed a milder flavor than those ripened with blue strains. In a later paper Morris[215] and his associates reported that of two white strains which they tested one was superior to the other. While to the individual who prefers a traditional strong flavored blue cheese, a mild-flavored one might be unsatisfactory, there are probably many who would prefer the milder type. For a more detailed review of current trends in blue cheese processing and marketing, the report of Scott[255] is recommended.

Camembert Cheese

Like Roquefort, Camembert is said to have originated in France, and like Roquefort it has several imitations. Named cheese varieties which are basically of the Camembert type are: Brie, Thenay, Troyes and Vendôme. As we have seen above, the principal role of the fungus in the ripening of Roquefort cheese is that of producing a highly unique flavor and odor through the hydrolysis of milk fats with the formation of various fatty acids, part of which latter is then metabolized by the fungus to characteristic methyl ketones. The principal and obvious action of the fungus in the ripening of Camembert type cheese is that of altering the texture of the cheese, although changes in odor and flavor also occur. Thom[303] was of the opinion that certain characteristic flavors are due to the action of lactic acid bacteria.

During the ripening process (which ordinarily lasts three or four weeks) temperature and humidity are closely controlled. During this period the texture of the curd is changed by the activity of the microorganisms which grow on the outside (rind) of the cheese. Ordinarily these organisms are *Penicillium camemberti,* its white variety (var. *rogeri*), *Oidium lactis* and several species of bacteria. The curd is softened to a smooth buttery consistency through the action of proteolytic enzymes produced by microorganisms growing on the rind. It is obvious that in the ripening of Camembert cheese, a pure culture of a single organism is not involved, and in all probability it really doesn't matter what species is involved as long as it produces the proper extracellular protease in sufficient quantity to bring about the desired texture change and does not impart any off odors or flavors. When *Penicillium roqueforti* occurs as a contaminant on Camembert cheese it imparts a bitter flavor, and *Penicillium brevicaule* and related varieties of fungi may impart a strong ammonia flavor. Thom has stated that the curd is not inoculated with *Penicillium camemberti* except when a new factory is being established, thus providing further evidence that a desired fungus may become the dominant member of a microbial population.

Aging and Flavoring of Meat

While beef is one of the preferred sources of protein in many areas of the world and at first glance its preparation for cooking would appear to be quite simple, it was learned many years ago that it is not ready for eating as soon as the beef animal is killed and butchered. While prehistoric man undoubtedly ripped off and ate chunks of raw meat from a just-killed animal, it probably didn't take him long to learn that for such meat to be tender and have good flavor it must be allowed to age. This flavor enhancement and the tenderizing which occur during the aging period are probably due to the action of various enzymes which are released from the cells of the meat once the tissues are dead. How much time is required for aging depends upon the environmental conditions maintained. If beef is held at a temperature near freezing (e.g., 34°F) to inhibit bacterial and mold growth, about 21 days are required for tenderization. However, an accelerated aging process may be used which requires only about 60 hours. In the accelerated process meat is held at a somewhat higher temperature (60-68°F). From the standpoint of economics, the accelerated method would seem most desirable; however, there are certain disadvantages associated with this method. When meat is aged at the higher temperature it may become contaminated with bacteria which can impart undesirable flavors and odors and in some instances completely spoil the meat. Microbial contaminants can be partly controlled by the use of ultraviolet radiation during the aging process. However, this practice also presents certain problems because ozone is frequently formed when ultraviolet radiation is employed, and ozone imparts rancidity to the meat. Furthermore, if the radiation is directed on the meat it may become "sunburned" and greater trim loss is incurred. Thus, it would appear that in spite of the more favorable economics of the accelerated method, the longer low temperature aging process is the best one to use.

Various individuals have noticed that if a steak or roast is stored at low (but not freezing) temperatures and that if certain molds develop on the meat, it is much more tender and has a better flavor. There have been (and still may be) several restaurants in the United States where steaks were always so treated, and at least one such establishment became rather famous in the Midwest for the fine quality of its aged steaks. The writer has processed steaks in this manner and has found them to be far superior, both in regard to flavor and tenderness, to steaks not so treated. When spores began to be formed on the surfaces of the steak it was regarded as aged and ready for cooking. The surface mycelia and spores were wiped from the meat with a clean cloth saturated with vinegar, and the steak then broiled in the usual manner.

Williams[337] has been issued a patent covering a process in which beef is aged in seven days by being deliberately exposed to inoculation with a mold. Briefly, the process is as follows: freshly killed beef is cooled at about 45°F for one and one-half to two days (long enough for rigor mortis to set in), and it is then hung in an enclosed space where the temperature is at least 45°F but less than 60°F, and the relative humidity is greater than 80%. A mold, *Thamnidium* sp. (class Phycomycetes, order Mucorales) is introduced into the atmosphere of the closed space and is periodically re-introduced every 3 to 12 hours. Eight different strains of Thamnidium were used and presumably all were satisfactory. According to Williams after three or four days the meat could be removed from the aging room and no trim was necessary. However, if left longer than four days some trim was necessary due to more extensive mold growth. The investigator expressed the opinion that tenderization occurred as a result of the penetration of mold hyphae into the meat, but it is doubtful that deep penetration would be achieved. It is far more probable that accelerated tenderization occurred as a result of the production of proteases by surface-growing mycelia. Williams was not specific as to how the mold was introduced into the atmosphere of the aging room, but it seems likely that spores were sprayed or blown into the air.

In a later patent Williams[338] described what

appears to be an attempt to develop a do-it-yourself, home tenderizing method for processing small individual cuts of meat such as steaks and roasts. In this second publication Williams mentions seventeen strains of *Thamnidium* but states that *Thamnidium elegans* is particularly desirable. Apparently it is proposed that the mold be cultured on a suitable medium enclosed in a very flat pillow made of mildew-proof fabric. When not in use this "pillow" is rolled and tied, but in use it is laid flat in a deep tray and covered with a layer of white pine sawdust moistened with some such material as citric acid (presumably to form a carbon source for the growth of *Thamnidium* and possibly to inhibit bacterial growth). A rack on which a steak or roast is laid is placed a short distance above the sawdust; a cover is then placed on the tray and the entire apparatus refrigerated at 32 to 50°F for 24 hours. The cut of meat is then turned over to expose the other side for 24 hours, after which time the meat is claimed to have been tenderized. These claims should be viewed with some skepticism, and how much tenderization would occur as a result of fungus activity would depend upon how much mold growth (if any) developed on the meat in a 48-hour period at a temperature as low as 32°F. Serious questions can be raised about the validity of this patent, since Williams redefines "thamnidium" to include "the class of molds referred to as Phycomycetes, Ascomycetes or Fungi Imperfecti or in the order Mucorales or in the family mucoraceae" (!). This is a most curious definition since *Thamnidium* has been a well-known genus ever since Link named it very early last century (*Berl. Mag. Naturf. Freunde,* 3, 31, 1809). Valid or not valid it is doubtful that the average housewife will leap at the opportunity to age cuts of meat in this fashion.

Oriental Fungus-Processed Foods

Waksman (*c.f.* Willaman[335]) has noted that not only can the Orient be distinguished from the Occident on the basis of geographical and ethnological grounds but also on the basis of the method by which starch is converted to fermentable sugar for the purpose of making beverage alcohol. This East-West contrast is also well-illustrated by the types of food which are processed by fungi. As we have just seen, in the Occident fungi are used as processing agents primarily in the processing of milk protein (i.e., cheese) whereas in the Orient fungi are used mainly to process soybeans, although a variety of other materials such as rice, wheat, peanuts, copra and fish are also so processed. The widespread use of soybeans can probably be explained on the basis of five events: (1) the long history of soybean culture in the Orient, (2) the serious over-population problem so prevalent in most areas of the Orient, (3) the general shortage of food, (4) the high protein content of soybeans, and (5) the advent of religions which excluded meat from the diet.

Although Shibasaki and Hesseltine[259] report that miso, a food prepared by fermenting mixtures of rice, soybeans and salt, has been prepared for over a thousand years in Japan, references to this and other fungus-fermented foods were not very abundant in Occidental literature until comparatively recent times. Church[61] described experiments in which Chinese ang-khak was prepared on a laboratory scale using *Monascus purpureus,* and in 1928 Dyson[87] published a short paper on shoyu (soy souce) and made passing reference to miso and tofu. Of especial interest is Dyson's accurate assessment of typical Occidental attitude toward mold-processed foods: "It is, however, with an element of surprise that we find staple food products prepared by moulds, the growth of which upon edible matter would be regarded by the Occidental as ample reason for assignment to the waste bin." This attitude must have changed somewhat during the next seven years, since Staley,[281] in an article entitled Soy Sauce Goes American, briefly described the production by scientifically controlled methods of a foodstuff long prepared by crude Oriental methods. Still another paper on soy sauce was published by Lockwood,[194] and Lockwood and Smith[195] not only presented an account of the manufacture of soy sauce but also gave very brief descriptions of the preparation of such food products as Miso (red and black), Sufu, Red Sufu, Chee-fan, Tsue-fan and Hon-fan. Further evidence that fungus-fermented foods have been used many years in the Orient is provided by the statement of these latter workers that the Chinese made soy sauce in ancient times as a household industry and that descriptions of the process for making soy sauce are found in books over 1500 years old. Occidental interest in Oriental fermented foods seems to have gained in intensity following the end of World War II. The reasons for this increased interest are varied and are probably to be sought among the following: (1) greater involvement of the United States in Asia during and after World War II, (2) relative abundance of P.L. 480 funds, (3) an increasing awareness of the many food problems created by ever-increasing populations, and (4) a partial alleviation of Occidental provincialism with respect to centuries-old eating habits. Following is an account of some of the principal fermented foods of the Orient. The account is by no means complete, since the entire subject needs rather critical study. There are a number of fermented foods which involve only the use of yeast and/or bacteria; such products, although they are undoubtedly important food items, are omitted from the present account.

Miso

This food, which is a paste of the consistency of peanut butter, is characterized by Lockwood and Smith[195] as the most important soybean food product in Japan, although Hesseltine[141] states that in terms of volume of production it is second to shoyu in Japan. The first mentioned investigators state that miso is made from soybeans, wheat or wheat flour, and rice in the proportions of 2:1:1, but Hesseltine and Wang[143] state that it is made from soybeans and rice (sometimes barley). In all probability there are a number of variations in ingredients as might reasonably be expected in any process that developed as a widespread household industry, but in the present account the process as reported by Hesseltine and Wang,[143] Shibasaki and Hesseltine[257-259] and Smith et al.[275] will be described. The process is a two-step one in which rice is molded with *Aspergillus oryzae* or *Aspergillus soyae* to prepare rice koji. This is then mixed with steamed, crushed soybeans; the mixture is then inoculated with a yeast *(Saccharomyces rouxii)* and allowed to ferment. According to Lockwood and Smith[195] the entire process requires about 10 to 40 days, the time largely depending upon the temperature at which the yeast fermentation is conducted. However, the rather carefully controlled process as outlined by Hesseltine and Wang[143] requires about 84 days, although Hesseltine[141] notes that the fermentation may last as long as two years. A brief summation of the process as described by these workers follows.

Polished rice is washed, soaked, the excess water drained, and it is then steamed. After steaming, the rice is cooled to 35°C, inoculated with *Aspergillus oryzae* and incubated for 50 hours at 27-28°C. At the end of this time the rice is permeated with mycelium, contains a variety of enzymes and is known as koji. Some hours before the koji is ready for use, soybeans are crushed in a roller mill to grits, washed and then soaked in water for 2.5 hours. The excess water is drained after which the grits are steamed for one hour under five pounds pressure and then cooled. They are then mixed with the koji and salt, inoculated with *Saccharomyces rouxii* (Hesseltine and Shibasaki[142]) and allowed to ferment for seven days at 28°C and two months at 35°C. Following this the mixture is ripened for two weeks at room temperature. It is then mashed and blended and the resulting product is known as miso. In their 1962 paper Shibasaki and Hesseltine[259] have described the above briefly described process in detail, and for the reader who desires greater details this paper should be consulted.

There is a variety of different misos and, hence, a variety of different names has been applied to this material. Different names might well be expected, since variations in ingredients, proportions of ingredients, fermentation time, etc. would all result in slightly different products. No attempt is made to seek out and list every name that has been applied; however, a few are here noted. Lockwood and Smith[195] refer to White Miso, Red Miso and Black Miso, and Nakano[217] classifies these on the basis of ingredients used as Mame, Kome and Mugi. Tamura et al.[300] list three types: Sendai, Edo and Mame, and Shibasaki and Hesseltine[259] speak of Sendai, which is dark in color, and Shinshu, which is white or yellow.

It is obvious that the matter of the nomenclature of miso needs some attention, and it is to be hoped that Stanton will sort out and clarify these names in his forthcoming book *Fermented Foods* (Academic Press, in preparation).

Miso is used primarily as a flavoring material, being added to soups and vegetables. However, Hesseltine and Wang[143] recommend spreading it thinly on cucumbers or substituting it for meat sauce on spaghetti. Tamura et al.[300] determined the amino acid content of three types of miso by microbiological means, and total amino acid contents of these food products are compared with the amino acid contents of casein and soybean meal in Table 5. From this table it is evident that the quantities and proportions of amino acids in miso compare quite favorably with those of casein and soybean meal. Apparently the high values of these amino acids are due to the presence of considerable quantities of free amino acids in miso. This could be due in part to the fungal digestion of some of the soybean proteins and rice proteins or could be due to the synthesis of certain amino acids in excess by the fungus.

Annual production of miso in Japan is quite high. Shibasaki and Hesseltine[259] estimate that about 974 metric tons are made annually and that about 13 million bushels of soybeans are used in the process. In addition, approximately 6.5 to 8 million bushels of rice would be used to prepare the koji for this volume of miso production.

Shoyu

In contrast to the miso just discussed, shoyu is a liquid food rather than a paste. Furthermore, this product, known as soy or soya sauce, is the Oriental food type most well known in the Occident. As early as 1928 Dyson[87] commented on the enormous scale on which shoyu is made and the great length of time required to make it (as long as two years). The Japanese Soy Sauce Brewers Association (Hesseltine[141]) states that the average annual per capita consumption in Japan is about three gallons. The process as described by Dyson[87] is rather primitive and the major steps in the operation are outlined in Figure 22. As may be seen from this figure,

TABLE 5

A Comparison of the Total Amino Acid Contents of Three Types of Miso with Casein and Soybean Meal

	Sendai	Edo	Mame	Casein	Soybean meal
lysine	7.4	8.6	9.5	8.0	6.6
histidine	1.4	1.5	3.3	3.0	2.5
arginine	8.7	8.8	6.8	4.0	7.0
aspartic acid	5.6	9.6	16.5	7.0	8.3
threonine	4.4	4.3	6.8	4.7	3.9
serine	6.0	8.6	11.2	6.7	5.6
glutamic acid	25.5	20.0	35.2	25.0	18.5
proline	5.8	6.4	10.7	11.0	5.0
glycine	7.4	6.4	12.6	2.5	3.8
alanine				3.0	4.5
methionine	1.2	2.3	3.7	3.5	1.1
cystine	**	**	**	1.0	1.2
valine	8.1	7.5	11.2	7.7	5.2
isoleucine	11.3	11.5	18.2	6.5	5.8
leucine	10.2	11.7	22.8	9.7	7.6
tyrosine	3.4	4.3	7.1	6.5	3.2
phenylalanine	5.9	5.5	6.0	5.9	4.8
tryptophan	1.0	0.9	2.0	1.2	1.2

Miso data based on the figures of Tamura et al.[300]

**Tamura et al. list no values for total cystine but report values of 0.01, 0.01 and 0.03 of free cystine, respectively, for Sendai, Edo and Mame.

the earlier method of preparing shoyu was a largely uncontrolled process. Dyson described shoyu prepared by this process as having a salty, meat-like flavor which he attributed to the presence of monosodium glutamate. In 1935, in a rather non-informative article, Staley[281] stated that the ancient industry of the Orient had been modernized by American manufacturers. However, in his rather cursory description of the process, he makes no mention of the use of any fungus; so it seems doubtful that he was discussing anything more than an imitation of the real Oriental soy sauce. More recently Hesseltine[141] has noted that a process involving the acid hydrolysis of soybean protein has been used in the United States but that the flavor of shoyu so prepared is inferior to that in which fungus enzymes are used for protein hydrolysis.

Lockwood[194] adapted the household method of the Chinese to pure culture techniques for making shoyu and used three microorganisms in the process: a fungus, *Aspergillus oryzae;* a yeast, *Hansenula subpelliculosa;* and a bacterium, *Lactobacillus delbreuckii.* On a laboratory scale his process required a minimum of 37 days and a maximum of 100 days and yielded 2.5 gallons of shoyu per 5 pounds of soybeans.

Since the early report of Dyson many improvements have been made in the shoyu process including the establishment of generally more hygienic procedures and more scientific control. The modern preparation of shoyu has been described by Yokotsuka,[346] and Hesseltine and Wang[143] have prepared a flow sheet for the manufacture of shoyu. Their flowsheet illustrates a somewhat different process from that outlined by Lockwood since they show a lactic acid bacterial fermentation followed by a yeast fermentation, whereas in Lockwood's scheme the yeast and bacteria were added simultaneously. In a process that developed as a household industry it would seem that there are probably many permissable variations in procedure.

For preparing the koji used in making shoyu, *Aspergillus oryzae* is commonly employed, although Hesseltine[141] reports that *Aspergillus soyae* is also used. Lockwood's earlier report that *Aspergillus flavus* is used is

FIGURE 22

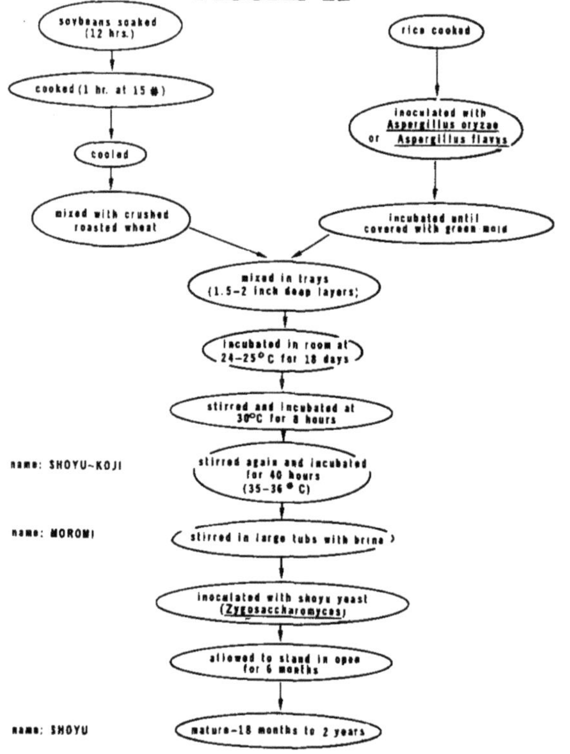

Processing steps used in the manufacture of shoyu—soy sauce.

somewhat puzzling, since some strains of this latter species are known to produce toxic materials (see under Toxic Substances in Fungi and Toxicoses). If, in fact, *Aspergillus flavus* was used, then it must have been a non-toxic strain; or if it was a toxin-producer, it did not produce toxins under the conditions under which shoyu was prepared. Dyson reported that the yeast employed was a *Zygosaccharomyces,* but Lockwood specified *Hansenula subpelliculosa.* However, Hesseltine[141] partially substantiates Dyson's earlier claim since he stated that in the latter phase of the shoyu fermentation, *Saccharomyces rouxii* and *Zygosaccharomyces* sp. are active. In all probability any one of a number of osmophilic yeasts might be used, and in a series of papers Onishi[224] has reported the results of his studies on the yeasts used in this fermentation.

Good shoyu contains large amounts of amino acids, one of the principal ones being glutamic acid. Therefore, Dyson[87] was probably correct in attributing much of the flavor of shoyu to monosodium glutamate. However, the large amount of sodium chloride, other amino acids and various of the fermentation end products all contribute to the flavor of this material. No attempt will be made here to enumerate all of the ways in which shoyu may be used, since this dark brown sauce is well-known in the Occident and has wide use as a flavoring agent in a great many different foods.

Dyson[87] and Lockwood and Smith[195] mention only one type of shoyu, but Hesseltine[141] briefly describes two. He states that in China more shoyu is the *tamari* type (more soybeans and less wheat are used), while in Japan over 90% is the *koikuchi* type (wheat and soybeans used in about equal amounts).

Writing in 1928 Dyson was impressed by the enormous quantities of shoyu that were manufactured annually. However, today, four decades later, he would be even more impressed, since Smith[274] stated that in Japan 1,061,000 bushels of soybeans and 7,778,000 equivalent bushels of soybean meal went into the manufacture of shoyu in 1962 alone. Since most Japanese shoyu is of the koikuchi type, an almost equal quantity of wheat would also have been used.

Hamanatto

Few references to this soybean product have been encountered in Western literature. Stanton and Wallbridge[282] make brief reference to the production of hamanatto in Japan, and one gets the impression from their Table 1 that a similar product called meitanza is prepared in China. Hesseltine[141] states that hamanatto is produced in a limited area in Japan by the fermentation of whole soybeans with strains of *Aspergillus oryzae,* but that it is expensive and very dark in color. His brief description (based on A. K. Smith. 1958. U.S. Dept. Agr. ARS-71-12) of the process is as follows: Soybeans are soaked in water for four hours after which they are steamed without pressure for ten hours. After cooling to 30°C they are inoculated with koji made from roasted wheat or barley and placed in trays. After 20 hours (when the beans are covered with green fungus) they are dried in the sun for one day. They are then placed in baskets with strips of ginger, covered with salt water and aged under pressure for 6 to 12 months. Upon removal from the brine the beans are red but they turn black when dried.

In a later paper Hesseltine and Wang,[143] on the basis of information supplied them by Dr. A. Kaneko of Nagoya University, give the following analyses of finished hamanatto: water, 38%; protein, 25%; carbohydrate, 25%; salt, *ca.* 13%. These workers have prepared a flow sheet of the process for making hamanatto, and, while it varies in some details, in the basic features it is much as described above. They state that this product is called *tu su* by the Chinese and *tao-si* by the Filipinos. No production figures for hamanatto are available, but apparently it is not produced in significant amounts.

Tempeh

Unlike miso, shoyu and hamanatto, this Indonesian food is made entirely from soybeans. Furthermore, it is not primarily a flavoring agent but is used as a main dish. One usually thinks of Oriental fungus-processed foods as involving *Aspergillus oryzae,* but in the preparation of tempeh a Phycomycete, *Rhizopus oligosporus,* is used. Also in contrast to the three soybean products just discussed, the production of tempeh is accomplished in a very short period of time. To be exact this product should be called tempeh *kedelee,* since a similar product called tempeh *bongkrek* is sometimes made from coconut meat. VanVeen and Mertens (*c.f.* VanVeen and Schaefer[315]) state that tempeh made from coconut meat may become poisonous but that tempeh kedelee never does. Since most tempeh is made from soybeans, only tempeh kedelee will be discussed here, and it will be referred to simply as tempeh.

Tempeh seems to have aroused considerable interest among Occidental investigators and as a result there is a considerably greater body of Western literature to be found relating to this food. VanVeen and Schaefer[315] describe it as easily digestible and quite palatable even to Occidentals which may in part explain the rather more extensive Occidental interest in this food product. Györky[132] stated that the nutritive value of one lot of freeze-dried tempeh was much higher than an unfermented soybean control and was equivalent to the nutritive value of skim milk. Autret and VanVeen[15] reported that the F.A.O. gave technical advice on tempeh manufacture in Southern Rhodesia and that results of large-scale experiments were very promising. Whether or not these experiments are continuing after the political change in Rhodesia is not known to this writer.

Steinkraus et al.[285] were among the first to produce tempeh on an experimental basis in the Occident. They isolated *Rhizopus oryzae* from crude Indonesian tempeh scrapings and used this organism to produce tempeh in small pans. Bacteria were also isolated from the tempeh scrapings but they found that bacteria were definitely undesirable in the tempeh fermentation, since they contribute off-odors. After cooked soybeans were inoculated and mold growth became rapid, the temperature rose and then gradually fell. Following this, sporulation occurred and some ammonia was produced due to protein breakdown. They pointed out that in small quantities neither sporulation nor ammonia spoiled the tempeh, a situation somewhat comparable to that found in Camembert cheese. These investigators followed the changes which occurred during the fermentation and found that while total nitrogen remained constant, soluble nitrogen rose from 0.5 to nearly 2.0%. The pH rose from 5.0 to 7.6, reducing substances showed overall decline, and fiber content increased from 3.7 in hulled beans to 5.85% in tempeh. The rise in fiber content was attributed to the development of mold mycelium. Upon sectioning soybeans that had been converted to tempeh, it was found that only slight penetration by mycelia had occurred; thus, the above noted changes were due primarily to the production of extracellular enzymes by the mycelia. These workers considered tempeh to be at its optimum organoleptically when the pH reached 6.5 and soluble solids were about 21%; at this point the tempeh was preserved by drying. Steinkraus and his associates obtained a yield of 72.5 g of tempeh (dry weight basis) for each 100 g of soybeans. They considered the following to be the most important points in tempeh production: (1) remove skins of soybeans, (2) adjust pH to 4.0-5.0 to prevent bacterial growth, (3) allow a sufficient supply of oxygen in the bean mass, (4) maintenance of a sufficiently high humidity to prevent drying, (5) control temperature so that it does not exceed 45°C during the fermentation. Wagenknecht et al.[318,319] studied the production of tempeh and found that during a three-day fermentation, *Rhizopus oryzae* hydrolyzed over one-third of the neutral fat. Rollofsen and Talens[245] studied the changes in thiamin, riboflavin and niacin content during the tempeh fermentation by *Rhizopus oryzae* and found that of the thiamin present in cooked cotyledons, about one-third was used up by the fungus. However, riboflavin increased by a factor of 3-5X, and niacin increased by a factor of 3.4X. In general Wang and Hesseltine[327] verified these changes in vitamin content in their investigation of the production of tempeh-like products through the fermentation of wheat with *Rhizopus oligosporus*. These investigators found that thiamin content of wheat tempeh was less than that of wheat, but that there were marked increases in both niacin and riboflavin. Hesseltine[141] states that there is also an increase in vitamin B_{12}.

Djien and Hesseltine[78] reported that work at the Northern Regional Research Laboratory has shown that tempeh can be made with at least four different species of *Rhizopus (stolonifer, oligosporus, oryzae* and *arrhizus)*. However, their species of choice appears to be a strain of *Rhizopus oligosporus,* and in this reviewer's laboratory the recommended strain has been used quite successfully. Apparently *Rhizopus oligosporus* is the species commonly used in Indonesia, since Boedijn[34] reported that this species can always be isolated from tempeh cakes. According to Djien and Hesseltine's description of the preparation of tempeh from yellow soybeans, the steps are as follows: (1) soak soybeans overnight, (2) remove seed coats, (3) cook for about 30 minutes in boiling water, (4) spread cooked beans on bamboo trays to cool and dry, (5) inoculate by mixing with tempeh from a previous batch, (6) wrap a handful of beans in a banana leaf to make a package about 1 x 5 x 10 cm, and (7) incubate at room temperature for one or two days. These workers stated that instead of being made in small packages in banana leaves, tempeh is sometimes also made in big cakes in split bamboo stems.

Martinelli and Hesseltine[203] developed procedures for tempeh production by pure culture procedures. The cooked, de-hulled, inoculated soybeans (*Rhizopus oligosporus, Rhizopus oryzae* or *Rhizopus arrhizus* were all satisfactory organisms) were packed in shallow wooden or metal trays with perforated bottoms, or in perforated plastic bags or plastic tubing. The soybeans were incubated at 31°C and good quality tempeh was produced in 24 hours or less. These same workers tried incubation temperatures of 20, 25, 28, 31, 37 and 44°C and soybeans failed to ferment only at 20 and 44°, although fastest fermentation occurred at 31-37°C. Thus, the tempeh fermentation can be conducted over a fairly wide range of temperatures. Perforated plastic bags were filled with inoculated soybeans and were then refrigerated or deep frozen. Upon removal from the freezer or refrigerator, 36 to 38 hours were required for frozen beans to ferment to tempeh but only 21 to 22 hours were required for refrigerated beans. Such a process has obvious advantages because it permits the preparation of a large batch which can be stored and then have portions fermented as needed.

Workers at both the Geneva Experiment Station and the Northern Regional Research Laboratory have investigated the large-scale production of this food product and for details of their work the original papers should be consulted. In the present instance the laboratory process as outlined by Hesseltine and Wang is described for the benefit of those who may wish to attempt to make small lots of tempeh. The steps to be followed are:

1. Dehulled soybean grits (not flour) are soaked in tapwater for two hours at 25°C.
2. After soaking, the grits are cooked for 30 minutes.
3. Cooked grits are then drained and cooled.
4. Cooled grits are inoculated with a spore suspension of *Rhizopus oligosporus* (NRRL 2710).
5. Inoculated grits are lightly packed in petri dishes and are incubated at 31°C for 20 to 24 hours.
6. After incubation the product is raw tempeh cake.

In the course of the incubation period, much of the soybean is solubilized by the fungus and the hyphae bind the particles together in a single mass (Figure 23). In a sense, the resultant product is similar to cheese although it does not have the uniform texture of cheese, since as may be seen in the accompanying figure, the identity of the individual soybean particles is not completely lost. Steinkraus et al.[285] have pointed out that during the fermentation, while total nitrogen remains the same, soluble nitrogen increases by a factor of about 4X. Thus, such water-soluble decomposition products as amino acids are produced in some abundance. Van Veen and Schaefer[315] stated that fermented soybeans are more digestible than unfermented ones, and Steinkraus et al.[283] suggest that they produce less flatulence. Sorenson and Hesseltine,[278] in their studies of carbon and nitrogen utilization by *Rhizopus oligosporus,* found that the principal soluble carbohydrates of soybean (stachyose, raffinose and sucrose) were not used as sole sources of carbon, but that various vegetable oils (cottonseed, soybean, peanut and rapeseed oils) supported good growth. In connection with this latter finding it will be recalled that Wagenknecht et al.[318,319] reported that in the course of the fermentation about one-third of the neutral fat of the soybean was hydrolyzed.

As noted earlier, tempeh is used as a main dish rather than as a flavoring agent for other foods. Raw tempeh is sliced and then fried or baked. Hesseltine[141] described tempeh fried in vegetable oil as delicious to eat when hot, and VanVenn and Schaefer[315] have pointed out that this food is quite palatable to Occidentals. In the reviewer's laboratory, quite acceptable tempeh has been prepared from a number of different soybean varieties, and while different individuals expressed different preferences for the different tempehs there was no general agreement as to which one was preferable. It would appear that studies on the nutritive value of tempeh are still being conducted, but enough information is presently at hand to indicate that its nutritional value is rather high. For example, Györky[132] fed rats on a 20%

FIGURE 23

Soybean tempeh (tempeh kedelee). Note texture of soybean "curd" after processing with *Rhizopus oligosporus*. Courtesy of C. W. Hesseltine, U.S.D.A.

tempeh diet, a 20% soybean diet and a 20% skim milk diet. Rats gained an average of 203 g on tempeh, 189 g on soybeans and 201 g on skim milk. In the rat feeding tests of Smith et al.[276] rats which were fed chip tempeh (soybeans mechanically dehulled and cracked into six to eight parts) or cotyledon tempeh (beans not cracked) did not show as great weight gains as those fed a diet containing 14% casein. However, when cotyledon tempeh was supplemented with 0.3% methionine, rats showed a slightly higher weight gain than those fed on the casein diet. A pancreatic hypertrophy factor known to be present in soybeans apparently was destroyed during the production of tempeh used in these feeding experiments, since no pancreatic hypertrophy was found in rats fed on the tempeh diets.

The writer has seen no production figures for tempeh, but enormous quantities of this material must be prepared. Production figures would at best be estimates because much of the tempeh is made in the home. Thus, Djien and Hesseltine[78] state that many thousands of people earn their living in this home industry in Indonesia.

Wang and Hesseltine[327] have prepared a material which they call "wheat tempeh," a food product prepared in much the same way as tempeh except that the substrate is wheat instead of soybeans. Although previous work had shown that *Rhizopus oryzae, Rhizopus arrhizus* and *Rhizopus oligosporus* were all suitable for the preparation of soybean tempeh, only the latter species was suitable for use in preparing the wheat product. The fermentation of wheat tempeh was complete in 20 hours when conducted at 31°C, and the writers described the taste of the cooked product as excellent. Although some sporulation of the fungus occurred at the edge of the cake, it was not noticeable in the product even if it was fermented as long as 43 hours. Thus, the production of wheat tempeh has a distinct advantage over the process of making soybean tempeh, since harvest time need not be rigidly controlled. Hesseltine[141] reported that at the Northern Regional Research Laboratory they were studying a whole new set of foods, tempeh-like products made with *Rhizopus oligosporus* from wheat, rye, barley and rice.

Both tempeh kedelee and the tempeh-like wheat product just described are of importance since they serve to make available more digestible and different types of food. However, their production adds nothing to the world supply of protein—it merely presents already existing protein in a somewhat different form. For that reason the work of Stanton and Wallbridge[282] should command more than passing interest. These investigators have developed a process for making a food product which, unlike the tempeh of Indonesia which is prepared from a high-protein substrate, is prepared from a primarily starchy substrate. The resultant tempeh-like product (Figure 24) has 6 to 7X the protein content of the original substrate.

One of the major problems in tropical and subtropical regions where manioc (cassava) is grown (Gray and AbouElSeoud[123]) as one of the dietary staples is the very high incidence of protein malnutrition. While this starchy root is high in carbohydrate its protein content is so low that in order to obtain one day's total requirement of protein (assuming that cassava protein is complete, which it is not), an individual would have to consume over twenty pounds of fresh cassava daily, an obvious impossibility for the average person. Unfortunately because of its ease of cultivation and relatively high-yielding properties, in many areas in underdeveloped countries cassava is one of the principal items of the diet. In such areas kwashiorkor is, of course, quite prevalent (Figure 25). This situation was most certainly not helped in Africa, where during the colonial period manioc cultivation was made compulsory in many African countries.

FIGURE 24

Fermented cassava—after five days growth of *Rhizopus arrhizus*. Below, deep-fried fermented cassava. Courtesy of W. R. Stanton and A. Wallbridge, Tropical Products Institute.

FIGURE 25

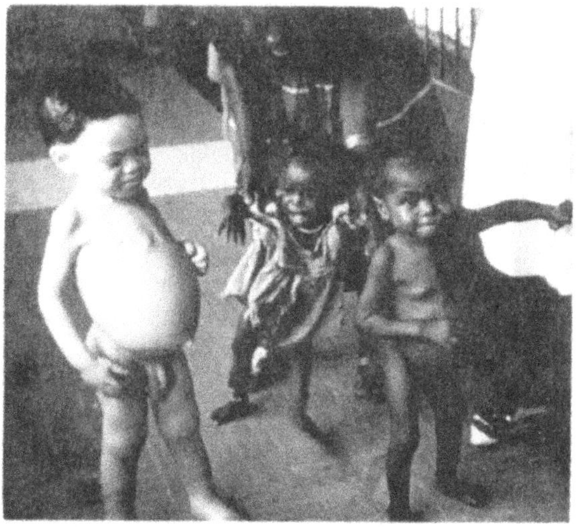

Sierra Leone child showing symptoms of extreme protein deficiency; note complete loss of skin pigmentation. Loss of hair pigment is concealed by use of black dye. From Gray and AbouElSeoud, *Econ. Bot.*, 20, 251, 1966. With permission.

In brief, the process consists of (1) making a stiff dough of cassava flour to which a simple nitrogen source has been added, (2) mixing fungus spores with the dough, (3) extruding the inoculated dough through a suitable machine into spaghetti-like rods, (4) breaking the rods into short lengths (Figure 26), (5) allowing them to ferment in small-mesh expanded aluminum trays. A temperature of about 30°C and a relative humidity of 95-97% were near optimum for the fermentation. Fungi tested for use in the process were *Rhizopus oligosporus, Rhizopus stolonifer, Rhizopus arrhizus, Rhizopus oryzae* and *Neurospora sitophila;* however, this last species proved unsatisfactory. While this process is obviously in the exploratory stage and needs further development and improvement, it appears to have considerable potential for placing more protein into the diet in cassava-eating areas. Curiously enough the concept of eating fungi with cassava apparently did not originate in the laboratories of Stanton and Wallbridge or Gray and AbouElSeoud (see page 271), since Hiscocks[145] reported that in Africa certain tribes prefer to eat cassava which is covered with a black mold.

FIGURE 26

Extruded cassava before fermentation and expanded aluminum trays used for the fermentation. Courtesy of W. R. Stanton and A. Wallbridge, Tropical Products Institute.

Ang-khak

This food product, also known as *ankak, ang-quac, ang-kak, anka, beni-koji, aga-koji, Chinese red rice* and *red rice* (Palo et al.[225]) is used as a food coloring agent rather than as a food by itself. Church[61] states that it is used to color Chinese cheese, and Hesseltine[141] reported that it is used in China, Taiwan and the Philippines for coloring cheese, fish and red wine, as well as other foods.

Ang-khak is prepared in certain provinces of China by culturing a red pigment-producing fungus on whole rice kernels until they are thoroughly permeated with the mycelia which produce the color. They are then dried and when ready for market can be easily powdered between the fingers, the powder imparting a red color to whatever food it is added to. The organism used is an Ascomycete, *Monascus purpureus*. Church[61] isolated the fungus from three Chinese red cheeses and from red rice, and much earlier Went[333] had reported that this fungus was responsible for the red rice of the Malay Islands and was used in Formosa for making a rice drink, *anchu*. Nishikawa[219] studied the pigment of ang-khak and found that the coloring material actually consisted of two pigments—a red pigment, monascorubrin ($C_{22}H_{24}O_5$) and a yellow pigment, (monascoflavin ($C_{17}H_{22}O_4$).

Church[61] noted that the Chinese were very secretive about the method of preparation of ang-khak but when she tested several strains of *Monascus purpureus* she found one (isolated from red cheese) with which she could produce a material quite similar to the Chinese product. She found that the amount of water used was quite critical and that if too much water was used, the typical red rice grains, each separate from the other, were not produced. Even during the fermentation the rice grains should not adhere together, but each grain should remain intact and discrete. Apparently the secret of successful ang-khak production is to have the rice grains just moist enough to permit the fungus to grow. Hesseltine[141] also cautions against the use of too much water, noting that at the time of inoculation the rice will seem rather dry. This latter investigator has briefly described the laboratory preparation of ang-khak, and the steps as he has described them are presented diagrammatically in Figure 27. Palo et al.[225] have shown that corn may be substituted for rice to produce this food coloring material, and in all probability a number of other grains could also be so employed.

Ontjom

Ontjom is a unique food of the East Indies

FIGURE 27

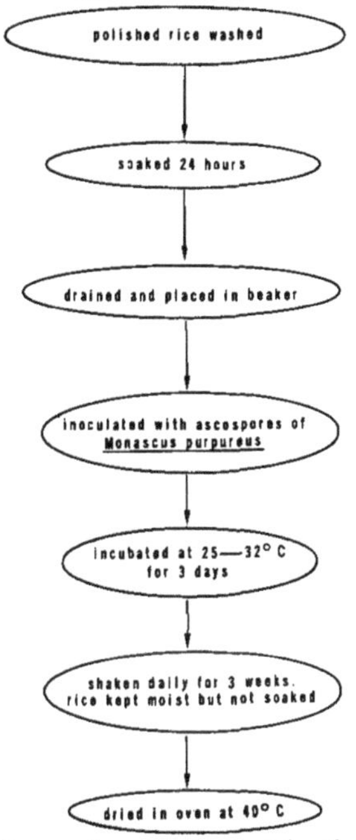

Processing steps used in the manufacture of ang-khak—Chinese Red Rice.

which is prepared with a fungus which is not used in the preparation of any other food product. While most Oriental fermented foods are prepared with a species of *Aspergillus* or *Rhizopus,* ontjom is fermented with *Neurospora sitophila.* Like tempeh, ontjom is produced in cakes and is used as a food in much the same way as tempeh, but unlike this soybean product, ontjom is produced from peanut press cake. Furthermore, because of the color of the fungus used, ontjom cakes are pink in color. A good account of the preparation of this food is presented by Ochse,[221] but Hesseltine and Wang[143] have presented a brief account more recently and the present discussion will be based upon this later paper.

In the production of oil from peanuts, most of the oil is pressed from the peanuts and a by-product of the process is the residual press cake. It is from the press cake that ontjom is made by the following steps:

1. Press cakes are broken up and soaked in water for 24 hours. During this step any remaining oil floats to the surface and is poured off.
2. Peanut residuum is washed several times and then steamed.
3. Steamed mass is then placed in small molds (3 x 10 x 20 cm).
4. Molds are placed in a bamboo frame covered with banana leaves.
5. Cakes in molds are inoculated by sprinkling with pink material (spores) from a previous fermentation.
6. Inoculated mass placed in a shady place and allowed to stand until the fungus has invaded the mass.
7. Cakes cut into pieces and eaten or taken to market.

Ontjom is always kept so that it is well aerated. It is eaten like tempeh and according to Hesseltine and Wang[143] fried ontjom tastes something like mince-meat. They also state that it is roasted, covered with boiling water and sugar and salt added or roasted, cut into pieces and covered with ginger sauce. Apparently some of the changes which take place during the fermentation of tempeh also occur in the fermentation of ontjom, since the strains of *Neurospora sitophila* which are used produce lipases and proteases.

There is a striking similarity between the processing of Roquefort cheese and Camembert cheese in the Occident and of tempeh and ontjom in the Orient. Although different fungi and different substrates are used, Roquefort cheese and tempeh are both inoculated throughout the mass, while ontjom is inoculated on the outside of the mass and Camembert cheese is left to chance inoculation on the outside of the mass.

Ontjom appears to be the only food made in quantity in which *Neurospora sitophila* is used as the fermenting fungus. Stanton and Wallbridge[282] found this species to be unsuitable for the fermentation of cassava; however, Steinkraus et al.[284] have used it to ferment soybeans. It is doubtful that this food would be especially appealing to the average Occidental, especially if he knew that the distinctive color was due to the presence of thousands of spores of a very common colored mold. For that reason it is suggested that if an attempt is ever made to introduce ontjom to Western tastes, it might be advisable to explore the possibility of using one of the albino mutants of *Neurospora crassa* as the fermenting fungus.

Sufu

Although the solid or paste type food products just discussed bear some superficial resemblance to cheese as known to the Occidental, it is sufu which most closely resembles this western food type. In fact, sufu is often called Chinese cheese, although it is not prepared from milk and Stanton and Wallbridge[282] go so far as to term it "true" Chinese cheese. The Chinese apparently have not devised as many "cheese" varieties as are commonly encountered in the western world. However, there are references to *red sufu,* a sufu colored with ang-khak in the course of its preparation. Like tempeh, the processing of sufu is accomplished with a Phycomycete, and Hesseltine[141] lists at least five species of mucoraceous fungi which have been isolated from sufu. However, in the work conducted at the Northern Regional Research Laboratory, *Actinomucor elegans* appears to be the organism of choice. From the flow sheet outlined by Hesseltine and Wang[143] sufu is made as follows:

1. Soybeans are washed, soaked in tap water for 5 to 6 hours at 25°C.
2. Soaked soybeans are ground with water and strained through cheese cloth.
3. Calcium sulfate or magnesium sulfate is added to the soybean milk that passed through the cheese cloth.
4. The mixture is then heated to boiling to coagulate the proteins and the resultant curd is pressed to remove water. At this stage the product is known as tofu.
5. The tofu is cut into small cubes which are sterilized for 10 minutes at 100°C in a hot air oven and placed in trays.
6. The cubes are inoculated with *Actinomucor elegans* and incubated at 20°C or lower for 3 to 7 days.
7. At the end of the fermentation period, when the cubes are covered with white mycelium, they are placed in a mixture of brine (2-5% NaCl) and red wine and aged for 40 to 60 days.

Hesseltine and Wang[143] report that sufu is widely used in China and was being prepared long before the Ching dynasty. These authors state that various materials can be used during the aging process, so there are probably a great many variations in taste, color and odor which superficially at least might be analogous to the seemingly endless variety of process cheeses now to be found on the American market.

Apparently tofu is often used directly as a food rather than being converted to sufu, and Wang[326] states that this has been one of the most important foods in the Orient for centuries. In a sense this would be the Oriental equivalent of curdled milk. For more specific details of the laboratory of pilot plant scale production of tofu the reports of Wang[326] or Smith et al.[277] are recommended.

Meitauza

In a protein-deficient area such as east Asia it would be unthinkable to discard any vegetable protein food. Therefore, the insoluble protein and other materials which are separated out from the "soybean milk" in the preparation of tofu are used to prepare a food product known as meitauza. These residual solids are pressed into cakes and allowed to ferment for 10 to 15 days. Shih (*c.f.* Hesseltine[141]) named the organism used in the fermentation *Mucor meitauza,* but Hesseltine states that this is a synonym for *Actinomucor elegans,* the organism which is used in the preparation of sufu. After a suitable fermentation time the cakes are sun-dried for a few hours and sold. They may be cooked alone in oil or cooked with vegetables. Apparently meitauza is made only in certain regions of China.

Ketjap

Although the principal soy sauce of the Orient is shoyu, another type of soy sauce is made in Indonesia. As described by Djien and Hesseltine[78] ketjap is made from black soybeans according to the scheme outlined in Figure 28.

As may be seen by comparing Figure 28 with Figure 22, the preparation of ketjap is a much less involved and a much shorter process than that used in the preparation of the better known shoyu.

Katsuobushi

In any densely populated area in underdeveloped countries, vegetable protein constitutes

FIGURE 28

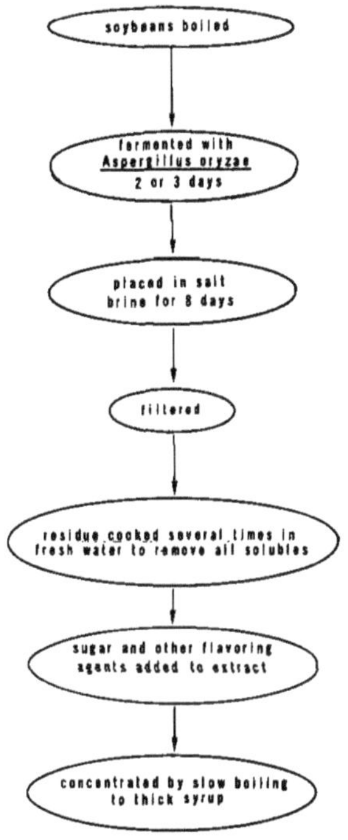

Processing steps used in the production of ketjap from black soybeans.

a much larger proportion of the diet than does animal protein. This may possibly be explained, at least in the Orient, on the basis that religions which exclude meat from the diet are more widespread there. However, this argument may be reversed and it certainly seems within the realm of possibility that such religions were initiated and gained impetus as a result of animal protein shortage. In spite of the fact that animal proteins are generally not as abundant in the Orient as in the more highly developed western countries, several fermented foods are prepared using animal (fish or shrimp) protein. Thus, Church[61] implied that the "freckled" codfish of China were speckled due to the growth of *Monascus purpureus,* the organism used in the preparation of red rice. This seems a rather minor use of microorganisms in food processing, but Hesseltine and Wang[143] make brief reference to fish paste (*bagoong*) which is a fermented product in which the involved microorganisms are not known, and fish sauce (*nuoc-mam*) which is a fermented fish product prepared through the use of halophilic bacteria. Saisithi et al.[248] state that 250,000,000 people in southeast Asia use fish sauce and list the following names for this product: *nam-pla,* Thailand and Laos; *nuoc-mam,* Cambodia and Vietnam; *patis,* Philippines; *ketjap-ikan,* Indonesia; and *ngapi,* Burma.

Hesseltine and Wang[143] speculate that the oldest food fermentations are fish fermentations but mention only one such product (*katsuobushi*) in the preparation of which in Japan a filamentous fungus is involved. Their description of the process is quite brief: "A bonito fish is divided into four or five pieces, cooked, and then fermented until dry." These investigators observed that members of the *Aspergillus glaucus* group were involved in the fermentation, a view expressed earlier by Thom and Church.[304] The finished product, which is dark, very hard and very dry, is shaved into ribbons which are used in flavoring other foods. The description of Hesseltine and Wang[143] of the apparatus used to shave katsuobushi into ribbons sounds very much like a description of the old-fashioned sauerkraut cutters which were in rather common use in the United States earlier this century. Kuninaka et al.[179] state that Kodama (*J. Tokyo Chem. Soc.,* 34, 751, 1913) found that the principal flavor component of katsuobushi was the salt of inosine monophosphate. Presumably other fish besides bonito could also be processed with an *Aspergillus* but whether or not they are is unknown to the reviewer.

Other Fungus-Fermented Foods

A variety of other products involving the use of fungi is also prepared in the Orient. Some of these such as koji and ragi are used in the preparation of alcoholic beverages of one type or another. Such materials play no part in the present discussion, but there are still other products, definitely food, in the preparation of which fungi play an important role. Although these are probably made in small amounts and are sometimes found only in rather restricted areas they deserve some mention here. For the present discussion the reviewer will depend largely upon the supplementary list in Hesseltine's[141] 1965 paper. For further details the reader should consult this supplement, since in most instances the original reference will be found there.

Chee-fan is a type of sufu made by salting small cubes of soybean curd, inoculating with a species of *Mucor* and allowing them to ferment for about seven days. After fermentation, the cubes are placed in yellow wine and mold of wine in a vessel and allowed to age for a year.

Fermented minchin is made by putting wheat gluten in a tightly covered container where after two or three weeks it is overgrown with both molds and bacteria. About ten per cent salt is then added and the mixture allowed to age for two weeks. The aged fermented minchin is then cut into thin strips and used as a condiment with other foods. It includes a mixture of molds such as *Paecilomyces, Aspergillus, Cladosporium, Fusarium, Syncephalastrum, Penicillium* and *Trichothecium.*

Fermented soybeans are prepared from black soybeans in China by a tray fermentation using a species of *Mucor* as the fermenting organism. After fermentation they are aged for six months or more in sealed earthenware jars. After aging the beans are seasoned with salt, spices and wine or whiskey and used as an appetizer.

Tao-cho is prepared from light colored soybeans. They are soaked, dehulled and boiled and then mixed with rice flour and roasted. When brown they are inoculated with an *Aspergillus* from teak leaves. They are then fermented for three days and sun-dried. The cakes are dipped in brine, arenga sugar and a paste of glutinous rice are added, and they are exposed to the sun for a month or more.

Tao-si is another food made from soybeans, but unlike tao-cho, the beans apparently are not dehulled. Soybeans are soaked and then boiled until soft. After cooling they are drained, the surfaces dried, and wheat flour is added until all beans are covered. They are then inoculated with *Aspergillus oryzae* and incubated for two or three days at which time the mass will be covered with mycelium. The mass is then placed in earthenware jars, salt brine is added, and the product is finished after about two months.

Taotjo is a condiment of the East Indies, also prepared primarily from soybeans. Boiled soybeans are mixed with roasted wheat meal or glutinous rice, wrapped in hibiscus leaves where they become inoculated with *Aspergillus oryzae.* After two or three days they are placed in brine and kept for several weeks, palm sugar being added to the brine at intervals.

POTENTIAL ADDITIONAL USES OF FUNGI AS FOOD

Although as early as 1798 Malthus spelled out the possibility of world-wide famine occurring as a result of continued human fecundity and is the one usually given credit for pointing to this potential problem of the future, he was by no means the first to recognize this developing problem. Benjamin Franklin foresaw this possibility in 1750 and thus antedated Malthus by nearly half a century, but even Franklin was a relatively late comer in this area, because famine had unquestionably been very much a part of the human scene throughout all recorded history. Thus, Biblical literature and early Egyptian literature make reference to famine. It may be recalled that Joseph ingratiated himself to Pharaoh by devising a plan whereby the "seven lean years" could be provided for in advance of their coming, probably the first recorded instance of such an attempt and possibly the only such attempt.

Nonetheless, modern concern about the possible dire results of unlimited population increase must be dated to the time of Malthus. However, Malthus' predictions seem to have made little impact at the time of their utterance, and it is only in relatively recent times that widespread concern has been exhibited regarding this problem in spite of the repeated warnings and predictions of demographers over the years. Lack of concern during Malthus' time is understandable, since world population at that time was such (*ca.* 750,000,000) that there were about 35 acres of arable land available per person, a more than adequate acreage to provide food for all, although by no means distributed equally. Hence, famine occurred periodically in various parts of the world and no general concern was expressed. However, as the science of demography developed, it has become more and more apparent to an increasingly greater number of people that widespread starvation among the human population is becoming a more distinct possibility. As a result more and more individuals are prophesying that a world-wide food shortage is imminent. In Table 6 are presented estimates of world population from 8000 B.C. to 2002 A.D. From these estimates it will be noted that by 2002 A.D. world population will reach at least six billion, twice the population of 1962. This estimate is based on a worldwide population increase of 1.8% annually, and if this increase remains constant, it is evident that, barring unforseeable cataclysmic events, by 2042 A.D. world population will be approximately twelve billion. The year 2042 A.D. may seem impossibly far in the future to many of us, but it should be remembered that a person born in 1970 would only reach the not uncommon age of 72 in 2042.

If for the moment we deal only with the projected population of 2002 A.D., it is evident that to maintain the population at that time at the same nutritional level as that enjoyed (or endured!) by the present world population will require that 1962 world food production be exactly doubled. Since it has been estimated that the inhabitants of two-thirds of the populated areas of the world now suffer protein malnutrition to a greater or lesser degree, the doubling of food production in the four decades following 1962 would merely result in making it possible for far greater absolute numbers of individuals to exist in a state of malnourishment. Perhaps at this point some thought should be given to the statement of Rock[243] that during the first four decades of the twentieth century, world food production was increased by only about 15 per cent. With that example in front of us we certainly have the right to hope, but do we have the right to believe that food production can be increased by 100 per cent by traditional agricultural means during the last four decades of the twentieth century?

In spite of alarming figures such as those presented above, there is still a very considerable difference of opinion as to how the impending problem can be solved. For a discussion of these various proposed solutions the work of Gray[115] should be consulted. The obvious and easy answer is to propose the regulation of population increase, but the achievement of such a solution seems very far in the future at this time. One needs but look at a country like India, where there are about a half-million births per week and peasant women drive birth control advocates out of the village with sticks, for confirmation of this view. In the meanwhile population continues to increase at an astounding rate. It is obvious that even if food production could be increased at the same rate as population increase, population regulation would ulti-

TABLE 6

Approximate Numbers of Individuals Living on the Earth at Various Times

Year	Approximate population	Average number of acres of arable land per person	Average decrease in acres of arable land/person/decade
8000 B.C.	10,000,000	2625	—
1 A.D.	250,000,000	100	3.15
1650 A.D.	500,000,000	52	1.29
1798 A.D.	750,000,000	35	3.40
1850 A.D.	1,000,000,000	26	1.80
1945 A.D.	2,000,000,000	13	2.88
1962 A.D.	3,000,000,000	9	2.35
2002 A.D.	6,000,000,000 ???	4	—

From Gray.[120]

mately have to be instituted or else overcrowding in itself would become a major problem. This latter problem is far in the future, however, because a few simple calculations reveal that at lease 600 years will have to elapse before there is "standing room only" on the earth. The prime objective at this time would appear to be that of increasing world food production by any or all means possible and at the same time mount a far-reaching educational program specifically designed to acquaint all people with the necessity for population regulation. Such a program would constitute the greatest educational endeavor of all time but could be justified on the grounds that for the first time in history all mankind is facing a common serious problem.

In spite of the seriousness of the situation, there still exist those who refuse to believe or at least to admit that all impending food problems cannot be solved by traditional forms of agriculture. Whether or not there is even an impending food shortage can be debated and the side one favors depends to a large extent upon whether one views food primarily as a source of energy or as a source of energy plus the all-important proteins that are necessary for the growth of the young and maintenance of the adult. In this writer's view it is highly improbable that there will be ever insufficient food production on the earth to meet the energy demands of the population for many years to come, since it has been calculated that the human population's energy requirements (allowing 3000 Calories per day per person) could be supplied by planting potatoes on about 2.5% of the world's arable land. It is also this writer's view that when we refer to an impending food problem we are referring specifically to an impending protein shortage problem.

In the United States the major efforts of most professional agriculturalists have always been devoted to increasing yield per unit area of land, often with little or no attention being paid to the nutritional quality of the crop being produced. Thus, it was possible to develop big potatoes with high water content which could produce greater tonnages per acre and thus net their growers greater returns on the market or in the government price paid for destroying(!), although the nutritional value of such potatoes is obviously less per unit weight (fresh basis) than is that of potatoes of lesser water content. A further example of the striving for greater yield per acre with utter disregard for nutritional value is provided by the development of hybrid corns which yield higher but have lower protein contents than the open-pollinated corn from which they were derived and which they have now largely supplanted.

With their phenomenal earlier successes in developing higher yielding varieties of crop plants there is little wonder that agriculturalists often came firmly to the belief that improved traditional agricultural methods could easily cope with vastly increased populations. One needs but look at any one of a number of emerging African countries to see the fallacy of this greater bulk of food philosophy. In many such countries one would be hard pressed to decide which caused the greatest harm, the introduction of syphilis or the introduction of hybrid corn. The introduction of hybrid corn into the agriculture of an uneducated population certainly has made the production of greater masses of edible bulk easier, but at the same time it has almost completely discouraged the cultivation of the more nutritious smaller grains traditionally grown in many localities.

As has been noted above, if one wishes to think only in terms of great masses of edible food materials with their high caloric content, there is no question but that traditional agricultural methods can provide adequate amounts of food for mankind for a great many years, rapid population increases notwithstanding. Unfortunately, however, this is not really the problem. The real problem is to provide increasing amounts of protein to keep pace with an increasing population. There are three points which must be kept in mind about this protein: (1) the total quantity produced must be such that about 65 g are available per person per day, (2) the protein must be of good quality; i.e., it must contain the essential amino acids in quantities approaching the proper amounts, and (3) it must be in concentrated enough form that an individual need not attempt to ingest almost impossible quantities of bulk to obtain his daily quota of protein. This reviewer is of the opinion that the demand for protein cannot be met by traditional agricultural methods in spite of many protestations to the contrary; in fact, it is not even being met today. In Table 7 are listed six crop plants, each of which constitutes a principal dietary staple in one or more protein deficient areas of the world. For example, west Africa is one of the areas in which manioc (cassava) is a food staple, and if the world supply of manioc produced in the 1962-63 season had been turned over to Nigeria alone, it would have supplied all of the protein that was needed by less than two-thirds of the population of that one country. If world production of manioc had been doubled (which it was not) in the seven years from 1962 to 1969, enough protein would have been produced to just supply the total protein needs of Nigeria in 1969. Even had this been done, protein deficiency would still have existed in that country for two reasons: (1) manioc protein is not a complete protein, and (2) to obtain their 65 g of protein, each person would have to eat over 13 pounds of fresh manioc root daily. Admittedly this example is an exaggerated one but it serves to again make quite clear the three basic aspects of protein which must always be considered: total quantity, quality and concentration. While quality and concentration are of extreme importance, it seems quite obvious that primary consideration must be given to the production of a sufficient supply to meet the requirements of the entire population, since incomplete proteins can be fortified and dilute proteins can be concentrated.

Man's ultimate source of proteins has always been the vegetable world, and unless a major breakthrough in chemistry is accomplished, the vegetable world will continue in this all-important role. For man this is an unfortunate accident in the evolution of the world of living organisms, since in the vast majority of instances the edible plant materials which man uses as food have low concentrations of proteins. The adoption in part of a carnivorous habit was a major step forward for the human species, since it enabled man to obtain his protein in concentrated form and thus allowed him time for other activities other than like a cow or an elephant the spending most of his waking hours eating hundreds of pounds of vegetable matter so that his body could sort out and utilize the protein it requires. One can be cynical about the manner in which man in general spends this leisure time afforded him, but an optimistic view would be that he will change and at least the potential is there. An alternative view would be that as a species man may have sealed his own doom when he adopted a partially carnivorous habit, because other animals are almost as inefficient as man in their metabolizing of protein, and the practice of eating great amounts of animal proteins involves the utilization of even greater amounts of vegetable protein. Thus, when the time comes (it may already have) when it is impossible to produce

TABLE 7

Six Common Crops Showing the Amounts Produced and Their Carbohydrate and Protein Contents

Crop	World production in 1962/63 (million tons)	carbohydrate (billion lb)	Total protein (billion lb)	Number of people who can be supplied with 52.2 lb protein/year(millions)
manioc	84.600	54.144	1.184	22.4
sugar beet	288.000	92.160	4.320	83.1
paddy rice	277.088	397.923	41.563	796.2
corn	243.600	292.320	34.094	653.1
yams & sweet potatoes	130.704	73.194	4.705	90.1
potatoes	294.324	112.434	11.772	225.5
Totals	1318.316	1022.175	97.638	1870.4

FIGURE 29

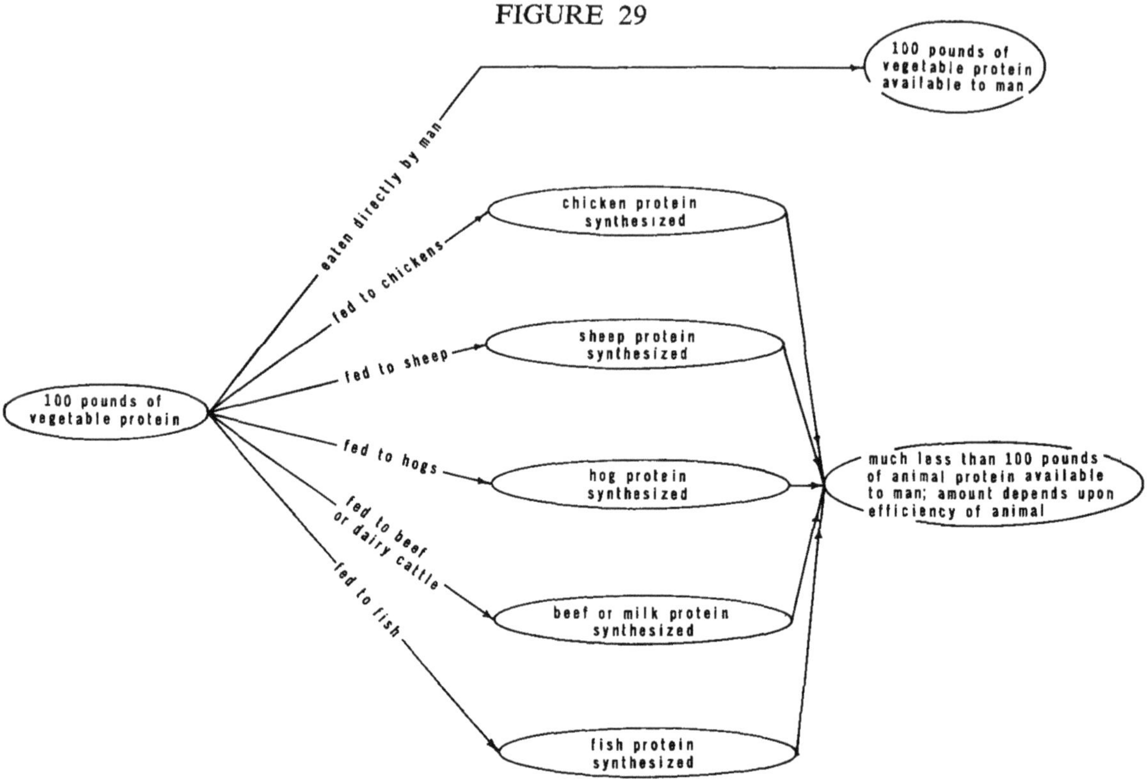

Possible ways in which man can utilize vegetable protein.

enough vegetable protein to produce the animal protein desired by man, he must choose between (1) adopting primarily vegetarian eating habits, (2) living in a chronic state of protein malnutrition, or (3) drifting back to the habits of Tshaka and his Zulu warriors who apparently often satisfied their animal protein needs by eating the captives which accrued to them in the course of their interminable "wars."

A simplified scheme of the possibilities which exist for man with regard to the handling of vegetable protein is presented in Figure 29. The limiting factor in all instances insofar as total protein supply is concerned is the maximum amount of vegetable protein which can be produced per unit area of land. The most efficient way to use this protein is, of course, for man to eat it directly and thus eliminate the protein losses due to the inefficiencies inherent in animal metabolism; the least efficient way is to produce beef protein from the vegetable protein since this process involves an 80 per cent loss in total protein. Thus, western man, who includes a considerable amount of meat and milk protein in his preferred diet, is faced with a dilemma which will become worse as populations continue to increase. "Highly civilized" western man will not take kindly either to a shift to a more vegetarian habit or a quiet suffering of malnutrition. Already this century we have witnessed several instances where population pressures have led to highly uncivilized acts (e.g., Italy's invasion of Ethiopia and Germany's "Drang nach Osten") and there is no reason to believe that such acts will not occur again when population pressures so dictate.

Traditional agriculture can help alleviate this situation because there are many arable lands which are not producing up to their capacity. However, there is a limit to the amount of vegetable material which can be produced on a particular piece of land as Mills pointed out quite clearly over a century and a half ago with his Law of Diminishing Increments. In some instances the substitution of high protein crops for high carbohydrate crops may help on a purely local basis, but again there are limitations to the amount of vegetable protein which can be produced per unit area of land as has been pointed out by Willcox[336] in his Inverse Yield Nitrogen Law. There are two reasonable alternatives: (1) man must carefully regulate population increase, or (2) he must avail himself of sources of protein which he does not now use. The first alternative above is quite patently the ultimate solution, but its institution will require many years, in the course of which the population will continue to increase. Therefore, it might be prudent to have a closer look at the second alternative.

Plants have the capability of performing a total synthesis of protein. That is, from carbohydrate and a few mineral salts they can first synthesize all of the amino acids which they require and then link these together in the proper proportions and sequences to form their proteins. Animals (including man) cannot perform such a total synthesis because they are unable to synthesize certain essential amino acids and, hence, cannot synthesize their own proteins unless they can obtain these amino acids (usually in the form of proteins previously synthesized by other living organisms) from other sources. Green plants have an especial advantage among living organisms because not only can they synthesize their amino acids from carbohydrates and a few minerals, but they also synthesize the carbohydrates. For this reason green plants are the ultimate source of all food for the animal kingdom. However, there are many thousands of non-green plants, which, while they are unable to synthesize carbohydrate, are able to perform a total synthesis of protein provided they are supplied with carbohydrate and a few mineral salts. It is the opinion of the present writer that vastly greater contributions to the world protein supply could be made if the carbohydrate-synthesizing capabilities of the green plants were combined with the protein-synthesizing capabilities of the non-green plants. For the latter, either fungi or bacteria could be used; however, because of the much greater number of species and the greater ease by which they can be harvested when grown in submerged liquid culture, fungi seem to have many advantages. The potential of such a process can be demonstrated with the following example. Because of their high protein content soybeans, which are about one-third protein, would seem to be a far more logical crop to plant for the production of protein than sugar cane which has a very low protein content. If protein quality is not considered, an individual's daily requirement could be satisfied by eating less than one-half a pound of soybeans, but he would have to eat over twenty pounds of sugar cane to obtain this same quantity of protein. Ironically, some sugar canes produce more total protein per acre than soybeans, but this protein is so dilute that it is of little or no value to man. However, a high-carbohydrate plant such as sugar cane produces vastly greater amounts of carbohydrate per acre than does soybean, so if some means could be devised to use this carbohydrate in the synthesis of protein a net gain in protein production per acre could be achieved if the conversion factor were a favorable one. Thus, Gray[116] has shown that if a fungus is used as the protein synthesizer, one acre of Peruvian hybrid cane can be the source of nearly three tons of protein as compared with the 800 pounds per acre of protein produced by an average yield of soybeans. It is with this concept of combining the carbohydrate-synthesizing capacities of the higher green plants with the protein-synthesizing capacities of the fungi that the present section is written.

Mycelia of Fleshy Fungi

In the earlier stages of the development of the field of industrial mycology, one of the major problems of scaling up to large volume fungus culture was that of obtaining sufficient inoculum. The efficiency of a large culture very frequently depended in large measure (with other conditions maintained at optimum) upon inoculation of a large volume of liquid medium at a great many different points. Hence it was necessary to be able to obtain many spores of the desired fungus quickly and easily. Thus we see that in the early development of the gluconic acid fermentation, which is conducted with *Penicillium chrysogenum,* it was not only necessary to devise a satisfactory fermentation medium but also a medium upon which spore production would predictably occur. At that period in the development of the field, species or strains of fungi which were sterile or which did not usually sporulate in culture were given little consideration as fungi of potential use in large scale pure culture operations. Among these neglected fungi were the fleshy Basidiomycetes and Ascomycetes, many of which do not produce asexual spores or, if so, not commonly or in the great abundance so common to most fungi of the mold type.

The requirement for copious spore production for large scale operation was eliminated; however, with the finding that spores were not really necessary and that good inoculum could be prepared by breaking up the assimilative hyphae of fungi (only those with septate hyphae) into fragments of a few viable cells each (Savage and VanderBrook[252]). Such fragmentation, which has now become routine procedure, is usually accomplished by aseptically transferring a liquid culture to a sterile Waring Blender cup and "blending" for a suitable period of time. Some mycelia can be fragmented by shaking with glass beads but the limitations of this procedure are such that it would be difficult to apply in the preparation of inoculum for a very large culture. Experience has shown that the proper blending speed and the length of the blending period vary with the organism being used, but proper time and speed can usually be readily established. With such a technique the need for producing large quantities of spores has been eliminated, and inocula which are more satisfactory than spore suspensions can be prepared, because the time period required for spore germination is eliminated. In our pilot plant work at Southern Illinois University we have found that sufficient inoculum for a 150 liter liquid culture can be prepared from a very small culture (50 ml).

With the development of this simple technique for preparing inoculum adequate for large volumes of medium, it was inevitable that attention be turned to a consideration of the possible use of fungi which hitherto could not be used. Therefore, it should have caused no great surprise when Humfeld[149] announced in *Science* in 1948 that he had succeeded in cultivating the mycelium of the Common Cultivated Mushroom, *Agaricus bisporus,* submerged in liquid medium. Actually, both Lambert[182] and Treschow[307] had previously shown that the mycelium of this species will grow submerged, but Humfeld apparently was the first to recognize that the production of such mycelium on a large scale might have commercial possibilities. In his first report Humfeld stated that he could obtain yields of mycelium of the order of 60% based on the weight of sugar utilized. Thus, from 100 pounds of sugar, 60 pounds (dry weight) of mycelium could be obtained, an Economic Coefficient* of 1.66, an example of the remarkable efficiency by which fungi may convert substrate carbon to tissue carbon. On a dry weight basis this mycelium had a considerably higher protein content than did the basidiocarp of the same species as may be seen from Table 8.

TABLE 8

A Comparison of the Protein, Fat and Ash Contents (Dry Weight Basis) of *Agaricus bisporus* Mycelia and Basidiocarps

	Basidiocarp[a] %	Mycelia[b] %
Protein (N × 6.25)	35.6	49.1
Fat (ether solubles)	2.3	3.1
Ash	10.2	8.1

[a] From McConnell and Esselen.[208]
[b] From Humfeld.[149]

Two significant points reside in Humfeld's accomplishment: (1) it provided means of obtaining virtually unlimited amounts of the assimilative phase of a mushroom which had been established through several centuries of use in the Occident as a perfectly safe and desirable addition to the human diet, and (2) it encouraged other investigators to attempt to culture other fleshy fungi of potential value. For many years a variety of fleshy fungi (both Basidiomycetes and Ascomycetes) has been known to be edible and tasty, but has been gathered for eating only as it

*Economic Coefficient (E.C.) may be defined in a variety of ways. In the present paper it is defined as the number of unit weights of sugar utilized in the production of one unit weight of fungus (dry weight basis).

formed large reproductive structures in nature, a seasonable and unpredictable occurrence. It now appeared probable that mycelia of many of these edible forms could be produced predictably in pure culture in unlimited quantities.

In his first work Humfeld used asparagus butt juice or press juice from pear waste as the main substrate, but later Humfeld and Sugihara[152] grew *Agaricus bisporus* in a completely synthetic and relatively simple medium of the following composition:

dextrose	50.0 g/liter
KH_2PO_4	0.87 g/liter
$MgHPO_4$	0.40 g/liter
$CaCl_2.2H_2$	0.37 g/liter
H_2SO_4 (2N)	5.7 ml
20% urea solution	15.0 ml
trace element solution	20.0 ml
$FeCl_3.6H_2O$	0.5 g
$MnCl_2.4H_2O$	0.36 g
$ZnCl_2$	0.20 g
$CuSO_4$	0.05 g
water to make 1 liter	

These investigators grew *Agaricus bisporus* in shake flasks, Fernbach fermenters and 20-liter fermenters and found that they could produce mycelia either without flavor or with a good mushroom flavor. To obtain the latter it was necessary to continue the culture for one or two days after all sugar was utilized, which suggested that flavor development was an autolytic process. After centrifuging and washing, the mycelium resembled yeast paste (20 g dry weight per 50 g of sugar; i.e., an E.C. of 2.5) and could be spray-dried, drum-dried, frozen or canned. They suggested that it be used in soups, as flavoring in sauces or as flavoring in meat dishes and spaghetti. In a later paper (Humfeld and Sugihara[153]) these same workers investigated the relation of nutrient requirements to both yield and flavor development. They found that requirements for nitrogen, phosphorus, potassium and sulfur for maximum flavor development were 1.5 to 2x, 6 to 8x, 1 to 3x and 4x, respectively, as great as for maximum yield.

Sugihara and Humfeld[292] next grew 23 fleshy fungi (representing 2 species of Ascomycetes and 18 species of Basidiomycetes) on solid media and in submerged culture and found that mycelia grown on solid media produced a typical mushroom odor which was not detectable if they were grown in submerged culture. Economic Coefficients varied from 1.6 to 3.3, but of the species tested only *Agaricus bisporus* and *Lepiota rachodes* had a pleasant flavor. Growth characteristics in submerged culture were not the same for all species; some produced mycelial pellets and others produced a dispersed growth with spores (probably oidia). They noted that the production of secondary spores is the chief characteristic of strains suitable for submerged growth. It is important to have knowledge of the growth characteristics because the harvesting of finely dispersed mycelia presents an altogether different problem than the harvesting of mycelial pellets. Humfeld [150,151] was issued two U.S. patents covering his process for growing mushroom mycelia.

In the meanwhile other investigators were exploring other possibilities. For example, Gray and Bushnell[126] screened 50 fungi, representing 12 species of Ascomycetes and 36 species of Basidiomycetes, in still culture and found that Economic Coefficients varied from 1.6 for *Boletus indecisus* (a Basidiomycete) to 43.0 for *Verpa conica* (an Ascomycete). Obviously an organism with an Economic Coefficient value as high as 43 would not be suitable for the economic production of large amounts of mycelia. They found that, in general, Economic Coefficients varied inversely with percentages of substrate carbon converted to mycelium carbon and respiratory carbon (carbon dioxide) but noted that the relationship does not represent a close correlation and they did not consider the use of E.C. value alone as an adequate basis for evaluation of the biosynthetic potentialities of a fungus.

For their studies Block et al.[28] used *Agaricus blazei,* an edible species that grows in Florida, since they found that it grows at higher temperature than *Agaricus bisporus.* Yield was very poor until they obtained a mutant strain that produced diffuse growth and secondary spores. These investigators cultured *Agaricus blazei* on various media containing such materials as citrus press water, orange juice, malt extract, nutrient broth or corn steep liquor and obtained most efficient conversion on medium with corn steep liquor. This species does not have a sharp pH optimum, since in the range of pH 3.5—7.5 yields were about the same; this can be a distinct advantage since it obviates the necessity for close pH control. Protein content of the mycelium was only 32.5% as compared with 43% found in a wild fruiting body of *Agaricus blazei.* Block and his associates expressed the opinion that the production of a true mushroom flavor remains the primary problem to be solved before commercial production of mushroom mycelium is practicable. In a later paper Block[26] reviewed the developments that had been made in the production of mushroom mycelia in submerged culture and noted that several times during the preceding decade commercial production had seemed imminent but that it had not yet materialized. He stated that there were no technical difficulties holding back the production of such a food and that the bottleneck seemed to be in the area of flavor development. Szuecs[294] was issued a patent for flavor enhancement of mushroom mycelium. The method involves the addition of 4 per cent salt and storage for 8 days at 4°C but it will not work if the mycelium is sterilized.

In England Eddy[89] cultured 28 fungi (of which 8 were strains of *Agaricus bisporus*) and noted that when *Coprinus comatus* was grown on the surface of liquid medium it had a strong mushroom-like flavor. For that reason further experiments were conducted to find a medium which would support good growth of this species. Highest yields (243mg/10 ml medium) were obtained on the following media: (1) 4% glucose, 0.09% urea, 20 ml-%dried autolyzed yeast, and (2) 6% glucose, 0.18% urea, 10 ml-% dried autolyzed yeast in 20 and 24 days, respectively. A lower yield was obtained on 2% glucose, 0.09% urea, 20 ml-% dried autolyzed yeast medium but efficiency of conversion of substrate carbon to tissue carbon was higher (i.e., E.C. value was lower). Eddy then cultured nine fungi, including *Coprinus comatus,* in a variety of ways and noted flavors and odors; in very few instances was either flavor or odor classified as mushroom-like. Eddy concluded that the production of a full mushroom flavor is connected with carpophore production, although some species under some cultural conditions will produce some flavor. He calculated that the cost of producing a pound of *Coprinus comatus* mycelium was about the same (11 shillings) as the market price of "mushroom powder,"* and expressed doubt that any market exists for tasteless mushroom mycelium, a doubt in which the reviewer concurs if it is proposed that such material be used as food for humans.

In Canada Reusser and his associates[240] screened ten fungi (seven agarics, one bolete, one morel and one *Xylaria*) on both molasses medium and waste sulfite liquor medium. Five of these produced satisfactory yields on molasses medium and of these five, three gave good yields on waste sulfite liquor. Of especial interest is the fact that *Tricholoma nudum* contained 38.7% protein when grown on sulfite liquor. On the basis of sugar utilized, 18.9 g of protein were produced per 100 g of sugar. Highest conversion of sugar to protein was obtained with *Morchella hybrida,* 19.7 g protein per 100 g sugar utilized. Protein content of *Tricholoma nudum* was much higher (54.4%) on molasses medium than on sulfite liquor, and on molasses medium four fungi besides this species were found to contain more than 40% protein. Since *Tricholoma nudum* had a pleasant flavor and aroma, Reusser et al.[241] studied its growth in relation to carbon sources, nitrogen sources, pH, phosphate concentration and aeration. They also studied its growth in industrial wastes such as beet molasses and sulfite liquor. They reported that this species has a pH optimum in the range of 3.5 to 4.5; however, if their data are used to calculate total protein in terms of yield in grams per liter (Figure 30) it will be seen that highest yields were obtained at 4.5 and 5.0. They found that tryptophane content was considerably higher than that reported for yeasts (Prescott and Dunn[231]) but that all other amino acids were present in about the same concentrations as in yeast. The mycelium was non-toxic to animals and had a B-vitamin content that was high enough to maintain

*Dried, powdered wild mushrooms such as *Boletus edulis.*

FIGURE 30

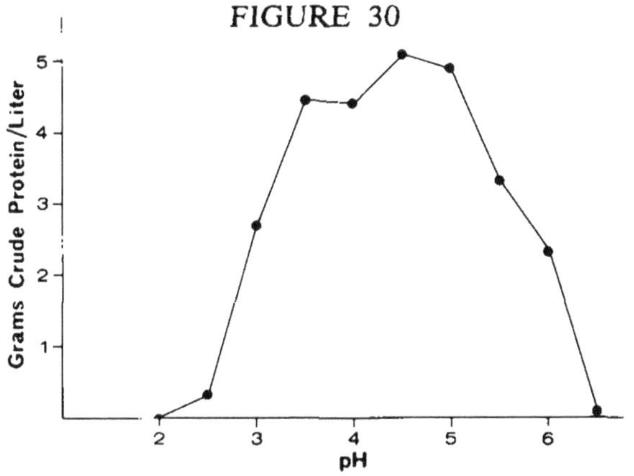

Relation of pH of medium to yield of protein synthesized by *Tricholoma nudum*. Based upon calculations of data of Reusser et al.[241]

normal growth in mice when used at a level of 5% by weight to replace vitamins in a synthetic diet. Cirillo et al.[62] also cultured various fleshy fungus mycelia on waste sulfite liquor and developed a process which they patented. These workers obtained highest yields with *Tricholoma nudum* and *Collybia velutipes*.

Like Block et al. and Reusser et al., Falanghe[92] attempted to produce mushroom mycelium on an industrial waste. Using the same ten fungi as those studied by Reusser and his associates, Falanghe found that only three (*Agaricus bisporus, Boletus indecisus* and *Tricholoma nudum*) could grow in a medium prepared from vinasse, a waste product from the distillation of fermented sugar cane juice. On the basis of percentage of protein, *Agaricus bisporus* was judged best; however, *Boletus indecisus* produced greater amounts of mycelium and hence was as efficient as *Agaricus bisporus* in terms of total protein synthesized. Both of these species formed mycelia which were readily separable by filtration, but the mycelium of *Tricholoma nudum* could not be easily separated by this harvesting procedure. Falanghe et al.[93] cultured various species of fungi on soybean whey, a waste product of the soybean protein isolation and concentration industry. Soybean whey contains about 11% of the soybean meal nitrogen but because of its low solids content it is usually discarded and has high potential for creating problems in sewage disposal systems. *Tricholoma nudum* and *Boletus indecisus* were found to be best suited for use in soybean whey, since their growth rates were faster, they produced more protein, and showed better utilization of whey solids. Mycelial growth of *Boletus indecisus* was always in the form of sphere-shaped colonies, but *Tricholoma nudum* failed to produce spheres if ammonium acetate was not added to the medium. Mycelial and protein yields of *Tricholoma nudum* greatly decreased if more than one per cent glucose was added to whey, but the yield of protein produced by *Boletus indecisus* was almost doubled when up to three per cent glucose was added. These results illustrate very strikingly the innate differences between species and should serve as an indication to any investigator that what may be optimum for one species may not be optimum for another. It is often possible, using a standard uniform set of conditions, to judge a species unsuitable for use whereas further study may reveal that a set of conditions can be found under which that species will perform quite satisfactorily. The highest yield in terms of grams of protein synthesized per liter was 6.3 produced by *Tricholoma nudum* in 12 days.

Jennison et al.[158] concerned themselves solely with the growth and nutrition of various wood-rotting Basidiomycetes, and Jennison et al.[159] studied the nutritive composition of mycelia of 17 different species (in 10 genera) of wood-rotting types. In the latter work all fungi were cultured in 6-liter portions of malt extract medium under forced aeration for seven days. They were then harvested, washed, freeze-dried and subjected to various analyses. Protein contents range from 21.7% in *Lentinus tigrinus* to 40.8% in *Polyporus tulipiferus* with an average of 30.6% for the 17 species. Fat contents varied from 2.3 in *Poria xantha* to 12.9% in *Peniophora gigantea* with an overall average of 7.0%. Mycelia were assayed for vitamin content, and tremendous variation from species to species was noted for each of the vitamins considered. For example, the thiamin content of *Hydnum pulcherrimum* was only 1.0μg per gram while in *Poria xantha* it was 56.1. Concentration ranges and averages for each of the seven vitamins are in Table 9. Uncooked mycelia were fed to mice and guinea pigs and none of the 17 species caused acute toxicity. The mycelium of one species

TABLE 9

Vitamin Concentrations in 17 Species of Wood-Destroying Basidiomycetes Showing Minimum, Maximum and Average Concentrations in Micrograms per Gram of Dry Mycelium

Vitamin	Minimum	Concentration Maximum	Average
thiamin	1.0	56.1	12.9
riboflavin	7.4	96.0	32.3
niacin	44.0	466.0	245.0
biotin	0.37	2.32	0.96
pantothenate	3.1	52.0	17.8
folic acid	0.34	9.1	4.35
pyridoxine	0.6	9.8	4.8

Based on data of Jennison et al.[159]

(Polyporus palustris) was substituted to the extent of 50% for laboratory rations and used in a one-month mouse-feeding experiment. The experimental animals gained weight and showed no symptoms of toxicity or dietary deficiency. However, mice that were fed only *Polyporus palustris* mycelium died after eight days showing symptoms of dietary deficiency. This latter type of trial can scarcely be termed as fair means of evaluation, since few if any single foods could be expected to supply all nutritional requirements. These workers found that the total amino acid content of the mycelium of *Polyporus palustris* was 32.4% of the dry weight. Seventeen amino acids (including those considered essential in human nutrition) were found. Of especial interest is the fact that lysine and threonine were found in concentrations of 2.1 and 1.8, respectively, since plant proteins are often low in these amino acids.

As noted in the Section on direct use of fungi as food, little success has ever been achieved in the deliberate cultivation and production of the fleshy ascocarps of various species of *Morchella.* Therefore, it was quite natural that attention would be turned to the possibility of producing morel mycelia that would have the characteristic flavor of the ascocarps. Szuecs[295,296] has been issued two United States patents dealing with the production of such mycelia; however, a considerable amount of the research and development work on morel mycelia appears to have been conducted at the Batelle Memorial Institute. The identity of the species of *Morchella* which Szuecs used is unknown. It is reported to have been isolated originally from an ascocarp collected in an orchard in Yonkers, New York, but no identification of the ascocarp was ever made. Curiously enough Szuecs chose to consider his morel to be *Morchella hortensis,* a species which has never been found in the United States!

In their brief review article of 1959, Robinson and Davidson[242] stated that production of morel mycelium (with the flavor of the morel ascocarp) had been engineered to full-scale production in 2000-gallon fermenters. They reported that Dr. Y. P. Chen had investigated the morel mycelium for possible toxic effects and concluded that a person would have to eat 60 pounds per day to have any ill effects. The product was then approved as a food by the Pure Food and Drug Administration of the United States.

Litchfield and Overbeck[191] reported on the submerged culture of *Morchella* species using food processing wastes as substrates, and Litchfield et al.[192] conducted a rather thorough study of the factors which affect the growth of morel mycelia. For their studies they used *Morchella crassipes, Morchella esculenta* and *Morchella hortensis,* but the major part of the work was conducted with the last-named species. Possibly this choice was made because of the slightly faster growth rate of this species which at 25°C gave maximum yields in four and one half days as compared with five and seven days, respectively, for *Morchella esculenta* and *Morchella crassipes.* Furthermore, using glucose, lactose or maltose as the carbon source, higher yields, higher protein contents and higher efficiencies of conversion of glucose to protein were obtained with *Morchella hortensis.* The highest percentage of protein found in this species when it was cultured on glucose was 34.8. In the pH range of 5.5 to 6.5 growth was constant, but it decreased at pH values above or below this range. Two types of medium (complex and synthetic) were used, the difference between them being that complex medium contained corn steep liquor but no added KH_2PO_4 or trace elements, while the synthetic medium contained both KH_2PO_4

FIGURE 31

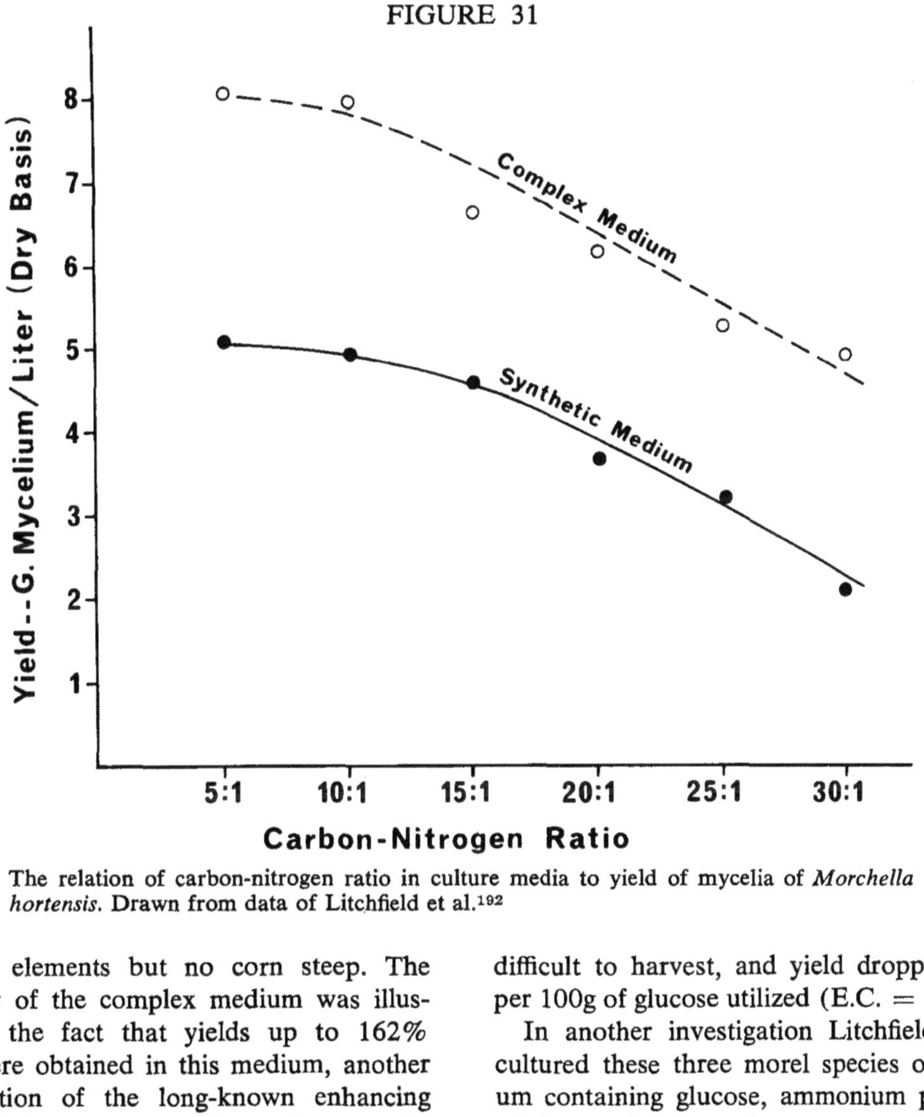

The relation of carbon-nitrogen ratio in culture media to yield of mycelia of *Morchella hortensis*. Drawn from data of Litchfield et al.[192]

and trace elements but no corn steep. The superiority of the complex medium was illustrated by the fact that yields up to 162% greater were obtained in this medium, another demonstration of the long-known enhancing effects which corn steep liquor has on the growth of fungi. Yield of mycelium per unit volume of medium varied inversely (Figure 31) with carbon-nitrogen ratio over the range of 5:1 to 30:1. These investigators pointed out that intense aeration was undesirable, since it not only led to decreased yields but also to different growth characteristics. At an aeration rate of 0.08 m mole of oxygen per liter per minute, firm mycelial pellets were formed and 47g of mycelium were obtained per 100g of glucose utilized (E.C. = 2.12), while at a rate of 0.15 m mole of oxygen per liter per minute the growth was somewhat filamentous but yield was the same. However, at an aeration rate of 0.20 m mole of oxygen per liter per minute, growth was filamentous, slimy and difficult to harvest, and yield dropped to 31g per 100g of glucose utilized (E.C. = 3.22).

In another investigation Litchfield et al.[193] cultured these three morel species on a medium containing glucose, ammonium phosphate, calcium carbonate and corn steep, and the harvested mycelia were subjected to various analyses which were compared with similar analyses made on a commercial product labeled "Morel Mushroom Flavoring." The commercial product contained 51% protein as compared with 22.8, 25.0 and 26.9 for *Morchella crassipes, Morchella esculenta* and *Morchella hortensis* mycelia, respectively, but contained less fat and less ash than did the mycelia of these three species. Amino acids were determined on protein hydrolysates, and 18 different amino acids were found and quantitated. No one of these three species could be said to be superior to the others in all respects; however, *Morchella crassipes* had higher contents of ten amino acids as compared with six for

Morchella esculenta and two for *Morchella hortensis*. When amino acid quantities of the commercial products were determined it was found that highest qualities of 12 amino acids occurred in this material. Glutamic acid, of course, was found to be present in greatest amount regardless of species, and rather significantly lysine, methionine, threonine and tryptophane were present in fair amounts in all four materials, although none of these four amino acids was reported to occur in as great amounts as they were in *Tricholoma nudum* (Reusser et al.[240,241]). The values reported for morel mycelia were similar to those reported for fungi by Anderson and Jackson,[7] and Litchfield and his associates concluded that morel mycelia have primary value as flavor supplements in foods and only incidental value in the contribution of certain amino acids. This appears to be an exceptionally shortsighted evaluation. The world has been well-favored with flavoring materials ever since the overland spice route was established through Samarkand to the Orient many centuries ago, and in a world in which protein deficiency is so widespread surely a better use than as a flavoring agent could be found for a material that is reputed to contain 51% protein. Reported amino acid values for any fungus mycelium should always be viewed with certain reservations. There is no suggestion here that any inaccuracies are involved; however, it should be borne in mind that the values found may merely reflect the particular set of environmental conditions under which a particular fungus was cultured. Over two decades ago Stokes and Gunness[287] showed that the amino acid content of the mycelium of *Penicillium notatum* could be varied by varying the cultural conditions. Thus it is not beyond the realm of possibility that by proper manipulation of cultural conditions the amino acid proportion of any fungus might be altered, and it is entirely conceivable that "tailor-made" fungus proteins may some day be a reality.

While other investigators (e.g., Moustafa[216]) have successfully cultured mycelia of various fleshy fungi, the above account seems sufficient to indicate that many such fungi can be cultured with good yields with no especial difficulty. Unfortunately, many of the investigators who have concerned themselves with the culture of the mycelia of edible mushrooms as well as other fleshy fungi seem to have been unduly preoccupied with the flavor and odor of the end product. This preoccupation seems to have stemmed from the idea that mycelia could and should be used in the same manner as mushrooms, principally as flavoring agents. The writer agrees that flavoring agents play an important role in food preparation; however, their overall direct contribution to man's nutritional status is now and always has been quite negligible. Furthermore, it is highly improbable that most people would easily accept mushroom mycelia as a substitute for mushrooms even though the flavors might be identical with those of mushrooms. Mushrooms have a distinctive appearance, a distinctive shape and a distinctive texture, and thus far no one has been able to duplicate these characteristics. Not only that, but little success seems to have been encountered in the area of flavor development. Litchfield et al.[193] state that mycelia of only *Agaricus bisporus, Lepiota rachodes, Coprinus comatus, Tricholoma nudum, Morchella hortensis, Morchella crassipes* and *Morchella esculenta* have been found to have a desirable odor and flavor. Szuecs[293] has patented a process for extracting essence of mycelium by steam distillation but it is quite doubtful that this process is in use. Stoller[288] expressed doubt that mushroom mycelium has ever been obtained with a strong mushroom flavor and notes that the mild flavor that it does have is lost in canning. With the well-developed common cultivated mushroom industry in this country and Europe and the rapidly developing industry in Taiwan, the production of truffles in France, and the production of Shiitake and the padi straw mushroom in the Orient, it would seem that any increased demand for mushrooms as flavoring agents could best be satisfied by further expansion of these already well-established industries. Development of a tasteless and odorless product might well be the primary objective in the production of mycelium and the resultant product viewed as an animal food which could be blended with other materials or flavored in such way as to make it palatable to the animal to which it is fed.

Mold Type of Mycelium

As may be seen in the preceding section, a great amount of research has been conducted by a number of investigators toward the end of developing methods for the mass production of the mycelia of a variety of fleshy fungi. Use of such mycelia has a number of advantages, the principal one being that the carpophores of many are known to be non-toxic and edible and hence one might reasonably expect the mycelia to also be non-toxic. There is also a psychological advantage in proposing the use of mycelia of fungi the fruiting bodies of which have been eaten by man for many years. Nonetheless there are certain disadvantages inherent in such fungi. In the first place, many grow slowly and their culture on a large scale would entail the involvement of expensive equipment for longer periods of time than might be economically feasible. In the second place, there are not nearly as great a number of species of fleshy fungi as there are of the more simple mold types; and in the third place, fleshy fungi do not normally produce an abundance of asexual spores. This latter disadvantage would certainly have no effect upon inoculum production as has been explained earlier, but it might be a major one in the establishment of a large-scale operation in which attempts were made to produce great quantities of mycelium in pure culture. It will be recalled that in the processing of Camembert cheese, where the same fungus (which is of the mold type) is used again and again in the same installation, it becomes unnecessary to inoculate the freshly-made cheeses; they become inoculated with the desired organism because through continued use it has become a dominant member of the microbial population. This can be a distinct advantage because under such conditions contamination cannot be considered contamination in the sense of another different organism gaining entry into the culture but merely consists of the gaining of additional inocula of the desired organism. For these reasons, it is believed that serious consideration should be given to the possible use of the mycelia of molds* for the production of large quantities of material of relatively high protein content. Such mycelia are not proposed for use as human foods except in situations in underdeveloped countries where a serious protein deficiency exists among a large segment of the population, and immediate remedial measures seem mandatory. On the contrary, it is proposed, just as it was proposed for mushroom mycelia, that such materials be used as animal feeds in order to provide means for increasing the amount of animal protein which can be produced per unit area of land.

The concept of using mold mycelia as food is most certainly not a new one. With the British naval blockade of Germany during World War I, this latter country was hard-pressed to find sufficient food for livestock, and Pringsheim and Lichtenstein[232] described the preparation of cattle feed from straw. This was an emergency measure, but it did make possible the preparation of a feed with about nine times the protein content of the original straw. Ammonium salts were added to straw which was then inoculated with a species of *Aspergillus*. After a suitable incubation period, the crude protein content was 8.0% as compared with an initial value of 0.9. Complete data are presented in Table 10, and from these data it would appear that the fungus grew largely at the expense of the pentosans and nitrogen-free extractables in the straw.

Skinner[270] was interested in determining if molds could synthesize such aromatic amino acids as tyrosine and tryptophane. He cultured three species of *Aspergillus*, two species of *Penicillium*, *Zygorrhynchus moelleri* and *Trichoderma königi* and on the basis of early studies selected *Penicillium flavo-glaucum* for feeding experiments. Mycelia were grown in still culture, autoclaved, filtered and dried and then incorporated as the protein source in rations for young rats. Both chemical tests and feeding experiments showed that this organism synthesized both tyrosine and tryptophane but that cystine was produced in such small amounts that it became the first limiting factor

*The term mold is here used in a broad sense to refer to all filamentous fungi which do not produce large (fleshy, leathery, wood or corky) sexual reproductive structures. Hence, molds would include many Phycomycetes, Ascomycetes, some Basidiomycetes and all of the Fungi Imperfecti.

TABLE 10

Partial Analysis of Straw Before and After Being Used as a Substrate for the Growth of *Aspergillus sp.*

Materials present	Untreated Straw (%)	"Fungus Straw" (%)
ash	9.1	14.0
ash (water soluble)	3.9	4.9
crude protein	0.9	8.0
ammonium sulfate	—	0.1
cellulose	50.3	50.7
pentosans	26.8	10.2
lignin	12.9	17.0
crude fiber	55.0	49.5
N-free extractables	35.0	28.4

From Pringsheim and Lichtenstein.[232]

in the growth of rats.

Takata[298] used mycelia of *Aspergillus oryzae;* Skinner et al.[271] used *Aspergillus fischeri, Aspergillus oryzae, Aspergillus sydowi* and *Penicillium chrysogenum;* and Woolley et al.[344] used *Aspergillus sydowi.* In each instance in feeding trials the fungal proteins were found to be inadequate, and on the basis of these reports Foster[105] stated that mold proteins are inadequate to satisfy the requirements of animals wholly. This curious and inaccurate generalization, based as it was on work with only a small fraction of a per cent of the many thousands of species of molds which are known, could have been avoided had the work of Vinson et al.[317] been considered. These investigators used mycelia of *Fusarium graminearum and Fusarium lini* in diets which they fed to mice. The first species proved inadequate, but normal growth, gestation and lactation occurred when mycelium of *Fusarium lini* was supplemented with thiamin.

Chastukhin[53] cultured mold mycelia on residual liquids from treacle, hydrolytic and sulfite alcohol manufacturing plants with promising results. He used a surface method of culturing fungi and reported yields of 100 to 150 g of dry mycelium per square meter of surface. Cattle fed readily on the mycelium, and because of its high protein content (32%) it was considered a good supplement to fodder low in protein. Practical realization of such a process proved impossible because of the need for large production areas and the great amount of labor involved. Thus, on the basis of the yields reported by Chastukhin, one acre of surface would be required to produce 428 pounds of protein. In a later paper Chastukhin et al[54] described their experiments in which 22 different mold type fungi were screened as to their abilities to grow on "treacle liquor"* from alcohol manufacturing plants. On the basis of these screening experiments four species (*Aspergillus oryzae, Aspergillus niger, Oidium lactis* and *Fusarium* sp.) were selected for further studies and average crude protein contents were found to be 23.95, 16.10, 28.80 and 26.80%, respectively, for these species when they were cultured on treacle liquor diluted 1:5. Protein contents were somewhat higher on treacle liquor diluted 1:1. It is impossible to calculate efficiencies from the data presented, but Chastukhin and his associates claimed that they had confirmed the possibility of utilizing fungi for the production of fodder proteins.

In 1960 Gray[115–120] and his associates initiated a program at the Ohio State University which was specifically designed to explore the possibility of producing protein through the culture of various members of the Fungi Imperfecti. The program was justified on the grounds that the world over-produces carbohydrates and under-produces protein. In the initial paper Gray[115] described the criteria for judging the suitability of a fungus for this type of use as follows:

1. The organism must be able to utilize nitrogen from an inorganic nitrogen salt as the sole source of carbon.

2. It must have the capacity to grow in submerged culture. This is a very important consideration, since we are concerned with the conservation of land area.

3. It must grow rapidly (we have arbitrarily set four days as the maximum acceptable length of the growth period).

*It seems possible that "treacle liquor" may be concentrated stillage from a plant in which alcohol is made by the fermentation of blackstrap or beet molasses.

4. It must efficiently convert substrate carbon to tissue carbon and at the same time accomplish a near theoretical conversion of inorganic nitrogen to protein nitrogen.

To the above criteria must be added a fifth: it must not produce any toxic substances. Of the 20 fungi first examined, 5 had Economic Coefficient values of 2.0 or lower. Of these five fungi *Epicoccum* sp. had an E.C. of 1.2 but had an exceptionally low protein content. However, *Heterocephalum aurantiacum* had an E.C. of 1.8 and protein content up to 35%. In preliminary feeding trials with mycelia of the latter species, adult mice maintained their weight on a diet consisting wholly of *Heterocephalum*. Some work was conducted with ground whole corn as the substrate for fungus growth and it was found that protein yield per acre could be doubled by using a fungus to convert the corn carbohydrate to fungus protein (Figure 32).

Gray[116] reported that 36 of the first 38 imperfect fungi examined were able to utilize inorganic nitrogen but that the same inorganic nitrogen source was not necessarily optimum for all fungi tested. Data on ten fungi are presented in Table 11 to illustrate this point.

When selected fungi were grown in an aerated large bottle culture (Figure 33) to test their ability to grow in submerged liquid culture they were found to vary widely with

FIGURE 32

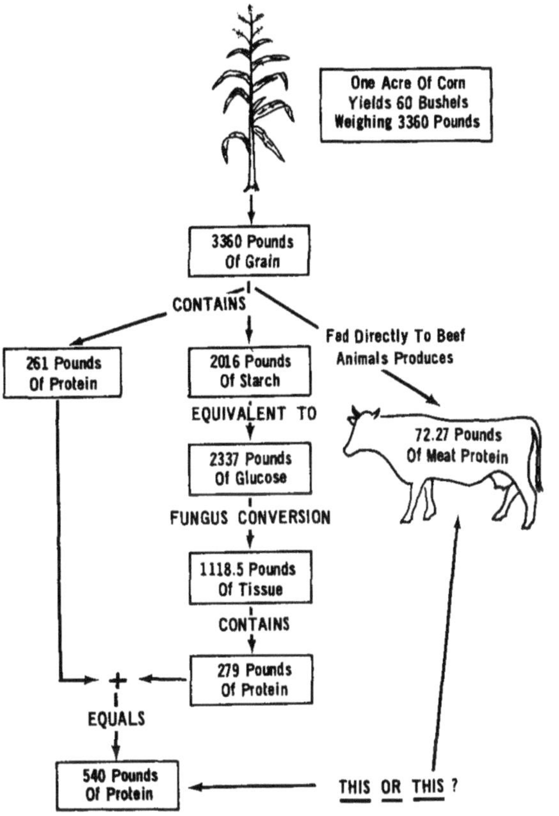

A comparison of the quantity of crude protein available from one acre of corn when fed directly to beef animals and when the corn carbohydrate is used in the synthesis of fungus protein. Redrawn from Gray, *Devel. Ind. Microbiol.*, 3, 63, 1962. With permission.

TABLE 11

Relation of Inorganic Nitrogen Source to Growth of Fungi Imperfecti. Still Cultures in 250 ml Flasks Incubated at 25° C for Six Days

Culture Number	Form Genus	Average Yield (mg/250 ml flask) Sodium nitrate	Ammonium nitrate	Ammonium chloride
I-99	*Sepedonium*	490	*498*	454
I-30	*Gliocladium*	414	*423*	249
I-36	*Cephalothecium*	415	*420*	183
I-11	*Myrothecium*	214	*299*	284
I-104	*Tritirachium*	96	*152*	122
I-75	*Cladosporium*	399	417	*461*
I-80	*Pullularia*	331	339	*449*
I-41	*Bispora*	53	53	*55*
I-73	*Trichurus*	*277*	233	109
I-19	*Speggazinia*	*203*	169	149

From Gray.[116]

FIGURE 33

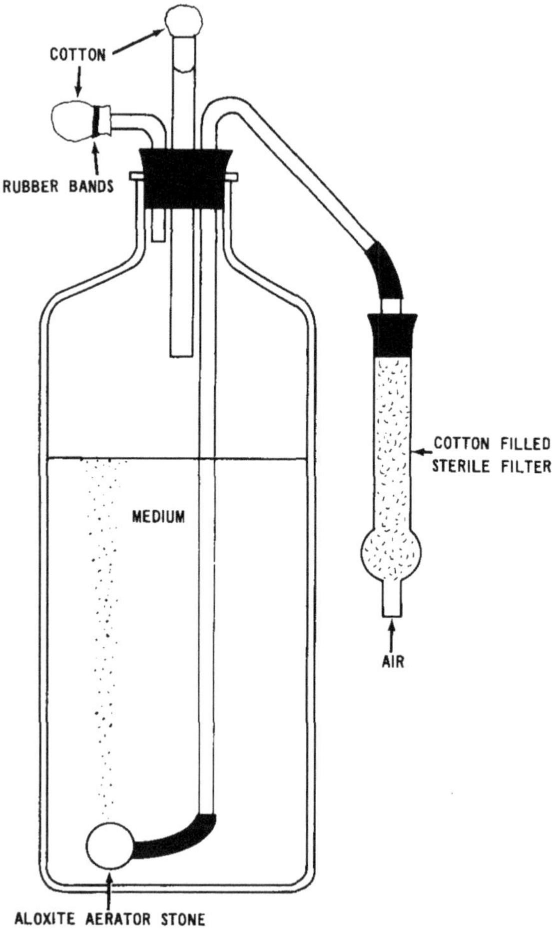

Large bottle method for growing fungi in submerged liquid culture.

respect to efficiency and per cent crude protein. In Table 12 are presented data derived from 11 different fungi.

From results presented in Table 12 it is evident that judgment based on efficiency of conversion of substrate carbon to tissue carbon is not necessarily the best way to select an organism for its ability to synthesize protein in large amounts. Thus, *Epicoccum* sp. can produce one unit weight of dried mycelium from 1.21 unit weights of glucose. However, the crude protein content of this mycelium is only 6.0%. On the contrary, *Heterocephalum,* which is less efficient (E.C. = 1.81), has a protein content of about 35%. The possibility of producing material with desired protein content (within limits) was illustrated by the results obtained (Figure 34) when *Cladosporium* sp. was cultured in medium in which the sole source of carbohydrate was ground sweet potatoes. Reference to this figure shows that all protein was synthesized by the second day but that mycelial yield continued to increase to the fourth day. A flow sheet for the production of fungal protein on a large scale was prepared (Figure 35), and it was pointed out that types of equipment necessary for such production were already standard items of use in industry.

Since large scale production of mold mycelium in submerged culture would require tremendous gallonages of water, Gray et al.[131]

TABLE 12
Growth of Fungi Imperfecti in Large Bottle (9 liters) Culture*

Culture Number	Form Genus	Nitrogen source	Mycelium dry wt/ bottle	G glucose required to produce 1 g	Per cent crude** protein
I–159	*Epicoccum*	NH_4NO_3	81.7g	1.21	6.00
I–9	*Heterocephalum*	NH_4NO_3	55.2	1.81	35.00
I–58	*Colletotrichum*	NH_4NO_3	55.2	1.81	11.56
I–80	*Pullularia*	NH_4Cl	51.8	1.93	——
I–83	*Cladosporium*	NH_4Cl	45.7	2.18	9.11
I–134	*Spicaria*	NH_4NO_3	41.1	2.43	25.25
I–30	*Gliocladium*	NH_4NO_3	35.7	2.80	18.92
I–73	*Trichurus*	NH_4NO_3	34.3	2.91	18.91
I–99	*Sepedonium*	NH_4NO_3	34.3	2.91	16.80
I–114	*Geomyces*	NH_4NO_3	22.4	4.46	25.92
I–29	*Helminthosporium*	NH_4NO_3	9.9	10.10	29.44

From Gray.[116]

*Four-day incubation at room temperature
**Kjeldahl nitrogen X 6.25

FIGURE 34

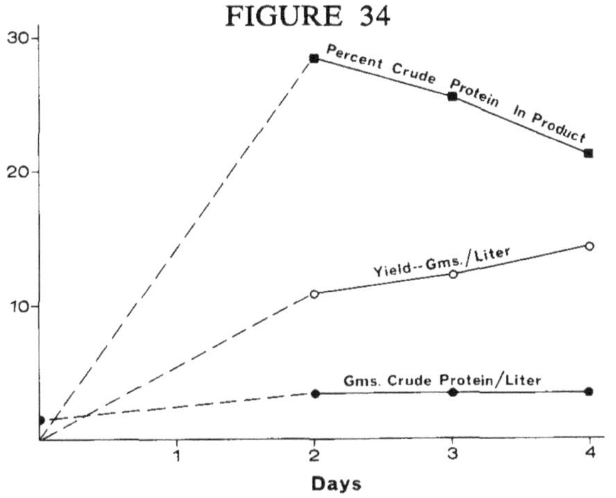

The relation between length of incubation period and protein synthesis by *Cladosporium* sp. on sweet potato medium. Redrawn from Gray.[116]

attempted to determine if culture medium could be prepared using sea water instead of fresh water. Twenty-three fungi (20 Fungi Imperfecti, 1 Phycomycete, 1 Ascomycete and 1 Basidiomycete) were cultured in media (containing no added magnesium) prepared with distilled water and sea water. Twenty-one of these fungi showed yield increases ranging from 7.6 to 178.5% when cultured in sea water medium. The beneficial action was considered to be due to magnesium ion added in sea water.

Gray et al[130] screened 175 different isolates of imperfect fungi in shake flask culture under constant conditions for four days and found wide variations in efficiency. The most efficient organism was *Rhizoctonia* sp. (O.S.U. No. 59) with an E.C. value of 1.30; the least efficient was *Botryosporium* sp. (O.S.U. No. I-189) with an E.C. value of 28.18. The distribution of E.C. values found for the 175 fungi is shown in Table 13, and from this table it is evident that over 50% of the fungi tested can produce one unit weight of mycelium from three or less unit weights of glucose. This does not necessarily mean that all test fungi which had higher E.C. values should not receive further consideration, since experience has shown that the efficiency of a fungus frequently can be improved by alteration of the culture conditions.

Both crude protein and extracted protein contents were quantitatively determined on 18 selected fungi and in most instances the crude protein percentage value was slightly higher than extracted protein percentage value. Highest extracted protein percentage of 32.0% was found in *Phoma* sp. (O.S.U. No. I-14). Lowest extracted protein content of 11.1% was recorded for *Rhizoctonia* sp. (O.S.U. No. 59).

Recognizing the fact that various carbohydrates vary in abundance and price from country to country and also that crude or waste carbohydrates are less expensive than purified carbohydrates, Gray and his co-workers explored the possibility of using such carbohydrates as substrates for the growth of imperfect fungi. The results of these investigations are presented in a series of papers in Economic Botany and will be cited specifically in connection with investigations of the suitability of various substrates. Thus, Gray and Abou-ElSeoud[122] used minced whole sweet potatoes as a source of carbohydrate and selected *Cladosporium* sp. as best suited for use on this substrate in submerged culture. The highest yield reported (in terms of grams of protein synthesized per liter of medium) was 5.87. From 100 g of sweet potato it was possible to produce 81.2 g of dried product (mycelium and unused sweet potato tissue) containing 31.6 g of protein. Since 100 g of sweet potato contain only 6.9 g of protein, total protein was increased by a factor greater than 4x. Such a process would be of great value in a country such as Egypt where high protein livestock feed is in short supply and the potential for producing sweet potatoes is very high.

TABLE 13

Distribution of Economic Coefficients of 175 Isolates of Fungi Imperfecti

Value of Economic Coefficients	Number of isolates	Percentage of total isolates
Less than 2.01	13	7.4
2.01 to 3.00	79	45.2
3.01 to 4.00	27	15.4
4.01 to 5.00	9	5.1
5.01 to 6.00	11	6.3
greater than 6.01	36	20.6

From Gray et al.[130]

FIGURE 35

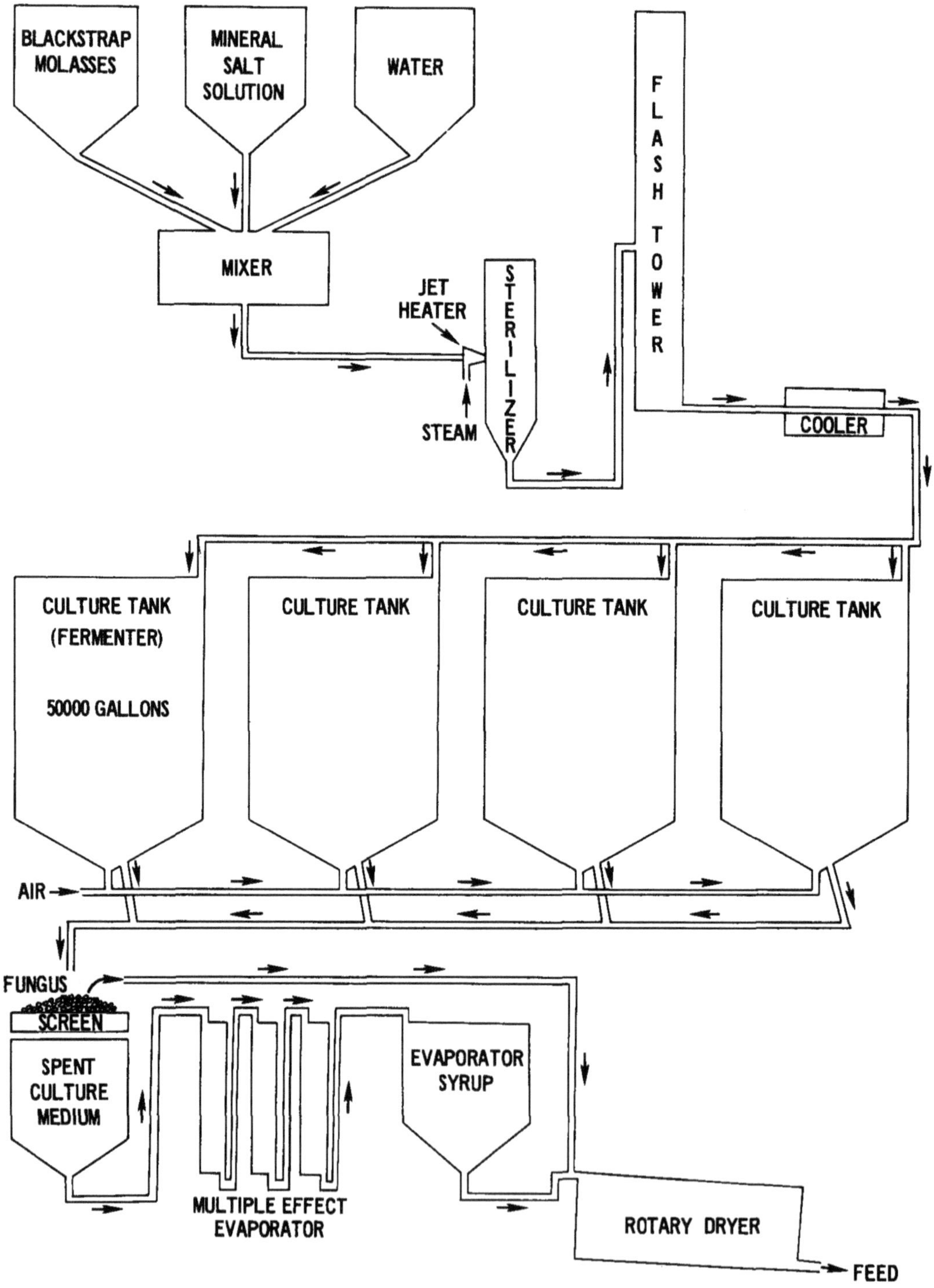

Proposed flow sheet for the production of livestock feed from fungus mycelium. Redrawn from Gray.[116]

FIGURE 36

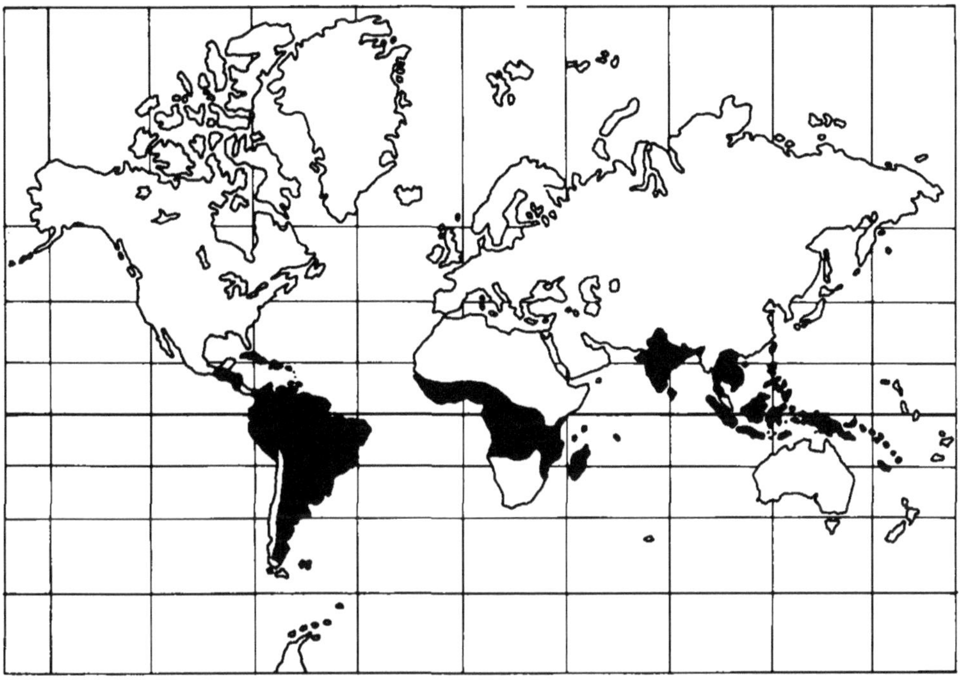

Manioc distribution. In the case of Africa, actual regions of manioc cultivation are indicated. In other countries where this crop is cultivated, the entire country is shaded. Redrawn from Gray, *Econ. Bot.*, 20, 251, 1966. With permission.

In the same manner Gray and Abou-ElSeoud[123] also explored the possibility of using manioc roots as a substrate. Manioc (also known as cassava, mandioca, aipum, yuca, cassada, tapioca, etc.) is one of the major root crops of the tropics although it may be grown in temperate regions such as the Southern United States. Manioc roots contain 32% carbohydrate and thus have high caloric value; however, they contain only 0.7% protein (Platt[230]) and, hence, in any area (Figure 36) where manioc forms a major dietary staple, widespread protein malnutrition may be expected. Several fungi were tested for their growth performance in medium in which manioc was the principal substrate, the *Cladosporium cladosporioides* (O.S.U. No. I-83) was chosen for use. Fresh manioc roots as well as the crude dried cassava flour so common in the markets of west Africa were added to media as the sole sources of carbohydrate, and while both could serve as carbon sources, more satisfactory results were obtained when fresh roots were used. It was found that by processing manioc in this manner a 5.7 fold increase in protein could be achieved in four days. When properly dried, fungus-processed manioc has an appearance very similar to the crude cassava flour found in African markets. Hence, if necessary it could be used as a diluent of unprocessed cassava flour and thus add measurably to the quantity of protein available in a manioc growing area.

While manioc is a very good source of carbohydrate to use in tropical or subtropical regions for conversion to protein using fungi as the conversion agents, its use in temperate regions would be quite limited. Therefore, Gray and AbouElSeoud[124] attempted to use minced whole sugar beets and beet shreds as the source of carbohydrate for the growth of various imperfect fungi in submerged culture. Beet shreds alone were found to be unsatisfactory as a carbon source but minced whole sugar beets were found to be an excellent carbon source. Several imperfect fungi were tested and again a species of *Cladosporium* (O.S.U. No. I-75) was chosen. Highest yields, both from the standpoint of total crude protein synthesized per 100 g of beet roots and percentage of protein, were obtained in medium containing 85 g of minced beets, 2ml corn steep liquor and 2 g of NH_4Cl per liter (Table 14).

From the results presented in Table 14 it is evident that with inorganic nitrogen source constant, higher yields (measured as protein synthesized per unit weight of beet roots) were obtained in medium of lower beet content. Whether it would be more economical to use a more dilute medium, however, would depend upon the cost and availability of beets as well as the cost and availability of water. Operating under those conditions which gave maximum yield, it was possible to increase the protein in beet roots by a factor of 2.6x. In the 1962-63 season, world production of sugar beets was 148,910,000 metric tons containing 1,116,825 metric tons of protein. Had these beets been processed with a fungus, the total protein available from the crop would have been 5,182,068 metric tons. The various possibilities for utilizing this beet crop other than in the manufacture of refined sugar are shown diagrammatically in Figure 37. In this figure it is assumed (1) that *Cladosporium* sp. is nontoxic to cattle, (2) that beef animals are 20% efficient in their conversion of vegetable protein, and (3) that 52.2 pounds of protein are required annually by each individual. To date, minced, fresh whole sugar beets have provided the best source of carbon for the growth of imperfect fungi. In some instances growth was complete in 30 hours.

Although rice is the most widely grown grain in the world today, in many areas such as India and Burma where it is the principal dietary staple, protein deficiency is quite prevalent. This is quite understandable since while rice consists of about 70% carbohydrates, its protein content is less than 10%. Thus, if a person eats only one pound of rice per day, he obtains less than 43 g of protein, about two-thirds of the amount that he requires, and protein deficiency symptoms may be expected to be manifested. Gray and Karve[127] attempted to increase the protein content of rice by substituting brown rice flour for glucose in the liquid medium described by Gray.[115] Fourteen imperfect fungi were tested and highest yields were obtained with *Trichoderma* sp. (O.S.U. No. I-193) and *Dactylium dendroides* (O.S.U. No.

FIGURE 37

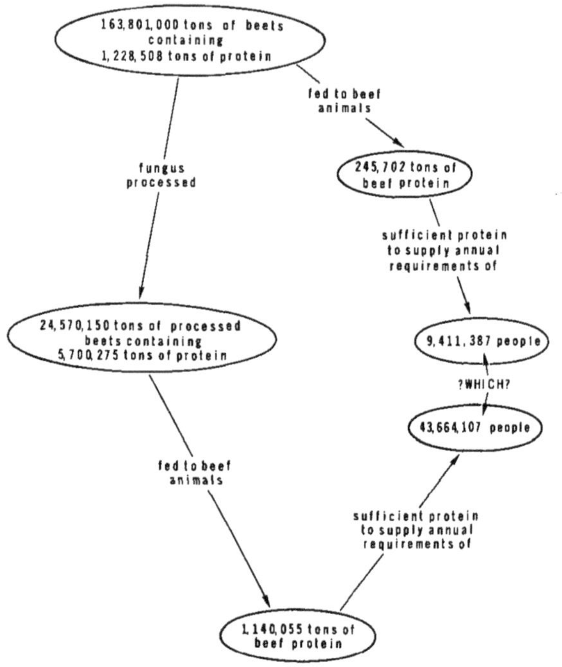

Alternate ways in which the world sugar beet crop could be used other than in the production of refined sugar. Based on total yield of the 1962/63 season.

TABLE 14

Relation of Growth of *Cladosporium* (I–75) to Sugar Beet Content and Inorganic Nitrogen (NH_4Cl) Content of Medium. Incubation Period: 72 hr; Initial pH: 5.0.

Sugar beet (g/liter)	NH_4Cl (g/liter)	Mycelial yield (g/liter)	Crude protein (%)	TCPL (g)	TCP/100g (g)
140	1.0	17.8	15.41	2.74	1.96
140	2.0	20.8	20.32	4.18	2.99
85	1.0	12.8	21.72	2.78	3.27
85	2.0	12.8	23.12	2.96	3.48

From Gray and AbouElSeoud.[124]

I-108). Highest yields were obtained on complete medium and in such medium it was possible using *Dactylium dendroides* to increase the total quantity of protein in rice by a factor of 2.29x. When *Trichoderma* sp. was used as the agent of synthesis, total protein quantity was increased by a factor of 2.85x. Properly dried fungus-processed brown rice flour did not differ appreciably in appearance from unprocessed rice flour, but its energy value was 11 to 17% higher than unprocessed flour. On the basis of a conservative estimate of protein being increased by a factor of 2.0x, it was calculated that if the 1962-63 rice crop had been used in a fungus conversion process, sufficient protein (on quantitative grounds) could have been synthesized to supply the annual requirements of 1.5 billion people, nearly half of the present world population. No feeding experiments have been conducted, so at present these data represent potential rather than reality.

Sugar cane is grown specifically for the production of more or less purified sugar and the extent of this production is illustrated by the fact that in the 1962-63 season 18,772,000 acres were devoted to this crop. It is significant to note that in all areas where the economy is based largely on sugar production, protein deficiency is present to a greater or lesser degree. For that reason Gray and Paugh[128] proposed that sugar cane be considered as a potential crop for the production of fungal protein. The more highly refined sugar is, the greater its cost, so the cheapest cane sugar available would be the sugar which occurs in freshly expressed cane juice. These investigators prepared media in which cane juice was added at such a level as to result in a 2% sugar solution. The *Cladosporium* (O.S.U. No. I-75) which had been used successfully in the processing of sweet potatoes and sugar beets was found to grow quite well in a medium based on cane juice. The best yield reported was 2.64 g total crude protein per liter which represents a trade of 7.57 unit weights of crude sugar for one unit weight of protein. On this basis the sugar in a 40 ton per acre yield of cane could be converted to over 1300 pounds of protein, and a high-yielding hybrid cane such as POJ2878 could yield over 3500 pounds of protein per acre if the sugar was used for the fungal synthesis of protein.

As has been noted earlier, supplying the total annual caloric requirements (estimated as averaging 3000 K cal/day/person) of the present world population presents no especial difficulties; however, countries such as India not only have a protein deficiency but also a deficiency of calories. In the synthesis of protein by fungi, some of the energy of the carbohydrate substrate is lost. For that reason it seemed advisable (Gray and Staff[129]) to measure the caloric values of the mycelia of a number of imperfect fungi that were being considered as possible synthesizers of protein. In spite of the tremendous number of investigations which have been concerned with various aspects of fungus metabolism (Raistrick et al.,[235] Tamiya,[299] Foster,[105] Rainbow and Rose,[234] etc.) determinations of energy values have been made on very few fungus species. Thus, Rubner[247] found that on combustion the mycelium of *Penicillium glaucum* yielded 4.753 K cal per gram. Terroine and Wurmser[302] and Molliard[213] both studied *Aspergillus niger* and obtained values of 4.8 and 5.9 K cal per gram, respectively. Tamiya[299] found that caloric value of *Aspergillus oryzae* mycelium varies with the type of compound used as the carbon source. Foster[105] stated that 5.0 K cal per gram is a generalized heat of combustion value of mold mycelium. In view of the paucity of reports on energy values of mycelium, Gray and Staff[129] made caloric determinations on mycelia of 100 different imperfect fungi selected from the 175 which Gray et al.[130] had studied. Caloric values varied from 3.658 K cal per gram in *Acremoniella* sp. (O.S.U. No. I-119) to 5.662 K cal per gram in *Geomyces* sp. (O.S.U. No. I-155). Much yet remains to be done in the area of investigation of the thermal efficiencies of fungi. Insofar as this writer has been able to determine, no one has as yet published results of a complete energy balance study of any fungus.

In spite of the fact that easily utilized carbohydrates (starches and sugars) are produced in great abundance, it would be far better from the standpoint of edible carbohydrate economy to utilize as a source of carbon and energy for mold growth some material which is abundant but has little use today as food. For that reason Chahal and Gray[43,44] considered the possibility of using wood pulp as the source of carbohydrate for the growth of fungi in liquid medium. Forty-two fungi (Ascomycetes, Basidiomycetes and Fungi Imperfecti) were screened for the purpose of determining which grew best in shake flask culture in liquid medium in which the principal carbon source was dried wood pulp ground through a 60-mesh screen. The four fungi which grew best and produced the greatest amounts of fungal protein were three imperfect fungi (*Rhizoctonia* sp., *Myrothecium verrucaria* and *Trichoderma* sp.) and one Ascomycete, the well-known cellulolytic *Chaetomium globosum*. Judged on the basis of total protein synthesized per unit volume of medium, *Rhizoctonia* sp. was best suited for growth on a cellulose-containing medium, although the mycelium of *Trichoderma* sp. contained a slightly higher percentage of protein. This fact was rather surprising in view of the fact that *Rhizoctonia* has never been reported to be a cellulose-destroying organism in nature and tends to lend substance to the optimistic view that if one searches far enough and widely enough, he can often find a fungus to perform the particular synthesis in which he is interested. Highest protein content reported was 17.8% and while at this time such a percentage is not considered adequate, it seems probable that further research might lead to the discovery of other fungi which can produce even greater quantities of protein from cellulosic substrates.

One mold, *Penicillium chrysogenum,* is widely cultured on a large scale, not for the production of mycelium *per se* but for the synthesis of penicillin, the mycelium usually being discarded. Pathak and Seshadri[228] proposed that such mycelium be used as animal food. Harvested mycelium was soaked in water for six hours, washed several times and dried in layers at 50°C. The dried mycelium was then substituted for soybean flour in mice rations (rations cooked), and at the end of a 29-day experimental period, the mice which had been fed mycelium showed weight gains comparable to those of control animals. These investigators noted that mice were very cautious with rations containing mycelium, but that once they became accustomed to it, consumption rate gradually increased.

In view of the above reports there is no question but that mycelia of a great many mold type mycelia can be produced in unlimited quantity and, insofar as total quantity of protein is concerned, can produce far more than is presently required by the human population. There are certainly no technical difficulties attached to their large-scale culture; experience in the production of penicillin and citric acid can be applied here to advantage. However, there are still a great many questions to which answers must be supplied. Extensive toxicity studies must be conducted as well as studies concerned with fungal protein quality. The industrial mycologist has demonstrated the feasibility of protein production by means of fungi, but the nutritionist, food technologist, toxicologist, protein chemist and chemical engineer must now enter the picture. In spite of the contributions of many investigators, only a fraction of a per cent of the fungi has actually been examined from the standpoint of their potential as producers of food.

The question of the economics of such a process is a perennial one and must be considered largely unanswered and unanswerable at this time. Estimates of costs have been and can still be made but the facts are that we will not know exact costs until large scale production has been accomplished. In the development of any microbial process earlier cost per unit of product, whatever the product may be, is quite high. However, as experience is gained and production is expanded, cost per unit decreases. There is no reason to believe that a process for producing fungal protein would prove to be an exceptional case. Estimates have been made in the United States which indicate that at this time fungal protein cannot be produced cheaply enough to be competitive with soybean protein. On the other hand, estimates have been made in Jamaica which indicate that fungal protein can be produced there at a lower cost per pound than soybean protein; both estimates may be correct. No generalizations may be made in this matter because many factors are involved and these factors differ from locale to locale. For example, when blackstrap molasses (an excellent source of sugar for the growth of fungi) were selling for $21.00 per ton in New Orleans, they were selling for $2.80 per ton on the Natal coast. On the other hand, inorganic salts of nitrogen were cheaper and more readily available in New Orleans than in Natal. Cheaper labor costs in Natal might be offset by the fact that in that province there is a much smaller pool of skilled labor. Cost of water, fuel, nitrogen source, labor and equipment all enter into the picture and will vary from country to country. The fact still remains that human population is increasing at an appalling rate, and protein malnutrition is keeping pace with it. Perhaps a more exact cost estimate could be made if some would put an accurate dollar and cents value on human health, human life and human dignity.

Myxomycete Plasmodia

The statements in the forthcoming short section must at this time be considered in the realm of pure speculation. In the preceding sections references have been made to various species in every class of fungi with one notable exception, the Myxomycetes. This class (plasmodial or acellular slime molds) is a unique group of organisms which have been of considerable interest to a relatively small group of biologists for many years. Consisting of about 425 species, their position in a system of classification is still in considerable doubt. In their assimilative phase these organisms exhibit many animal-like characteristics, but when they enter the reproductive stage, definitely fungus-like spore-bearing structures are formed. The first intensive studies on the group were conducted by the great German mycologist, Anton de Bary.[18-20] DeBary chose to regard the plasmodial slime molds as more animal-like than plant-like in their affinities and usually applied the name Mycetozoa rather than Myxomycetes to the group. However, the majority of workers today seem to prefer the name Myxomycetes, and except for certain biochemical and biophysical studies, a consideration of these organisms has largely been in the hands of mycologists. The most recent treatment of the general biology of these organisms is that of Gray and Alexopoulos.[125]

The structure of potential importance in the present discussion is the assimilative structure, the plasmodium, which is a naked, free-living, multinucleate, motile mass of protoplasm. Because they have no cell walls, plasmodia were early recognized as excellent sources of material for use in studying protoplasm and many of the early studies on the chemical and physical properties of protoplasm were conducted with such material. More recently they have become quite popular organisms for the study of the biochemical reactions involved in morphogenesis.

Although Myxomycetes had been cultured in the laboratory earlier by a number of investigators, the first major contribution in this area was made by Howard,[148] who devised a method for culturing one species, *Physarum polycephalum,* in such manner that large quantities of plasmodium could be obtained. This method was improved a few years thereafter by Camp[41] who devised a moist chamber culture method which made even larger quantities of plasmodium of *Physarum polycephalum* available. In neither instance (as was true of previously-used culture methods for this and other species) was it possible to grow the plasmodium in pure culture. Other investigators cultured other species (but never very successfully in pure culture) until by 1963 only about 10 to 15% of the known species (about 425 species total) had been reported to have been cultured by various workers. Of these, only about 11 species were reported to have been obtained in axenic culture.

The first major breakthrough in the area of axenic culture of Myxomycetes was made by Daniel and Rusch[71] and Daniel et al.[69] who reported the pure culture of *Physarum polycephalum* in shake flasks in a partially defined liquid medium. In such culture the plasmodia were not the typical fan-shaped structures so evident when the organism is growing on a solid substrate. On the contrary, they formed small, nearly spherical microplasmodia (Figure 38). Growth occurred only in the presence of an unknown factor present in chick-embryo extract, foetal-calf serum and foetal-calf erythrocyte haemolysate. Kelley et al.[166] and Daniel et al.[70] showed that the chick embyro ex-

FIGURE 38

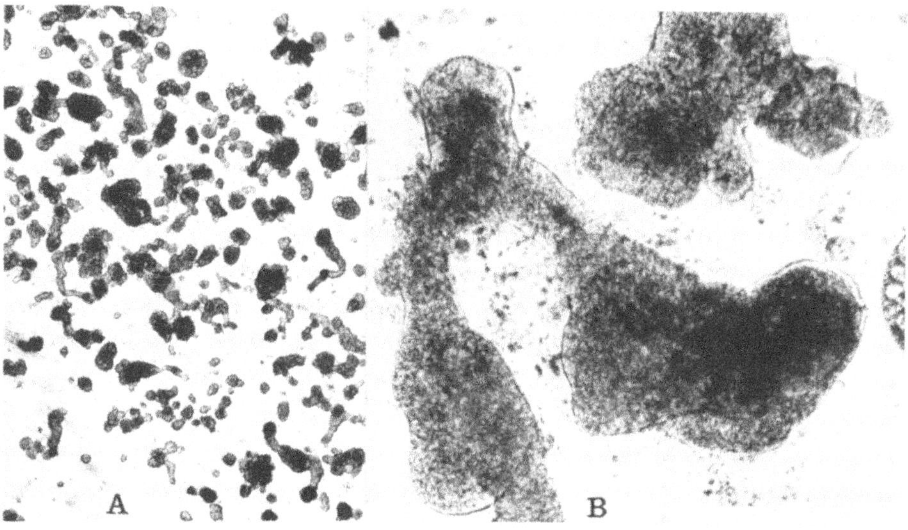

Microplasmodia of a myxomycete, *Physarum polycephalum,* grown in submerged culture; A—x12, B—x70. Courtesy of H. P. Rusch.

tract requirement could be met with hematin or certain hemo-proteins. In this latter paper it was established that in addition to hematin, the plasmodium of this species had absolute requirements for D- or L-methionine, biotin and thiamin. More recently, Daniel and Babcock[68] showed that S-methyl-L-cysteine, 2-hydroxy-4-methiol butyric acid, S-adenosyl-L-methionine and methionine peptides would support growth when substituted for methionine.

Two additional species of Myxomycetes have recently been obtained in pure culture in chemically defined minimal media by Henney and Lynch.[140] Like *Physarum polycephalum* both had an absolute requirement for hematin. *Physarum flavicomum* required methionine, glycine and arginine as did *Physarum rigidum*; in addition, the latter species also required valine. Henney and Lynch reported that with all three species, homocysteine thiolactone could be used to replace methionine.

Brewer et al.[37] first reported the large-scale culture of a myxomycete. These workers cultivated *Physarum polycephalum* in shake flasks, 2-liter bottle cultures, 15-liter cultures in 30-liter laboratory fermenters, and 100-liter cultures in a 50-gallon stainless steel fermenter. In the 15-liter cultures, yields of 10.5 g (dry weight) of plasmodia per liter of culture medium were obtained in four to five days. A similar yield was obtained in a 100-liter culture but growth was slower and ten days were required to attain such a yield in the larger culture.

The point of real interest about myxomycete plasmodia in the present discussion is that these organisms are not bounded by a rigid cellulose or chitinous cell wall but consist primarily of protoplasm, a considerable portion of which is protein. The production of microbial protein for food purposes has been advocated by a variety of workers, and various types of microorganisms have at one time or another been suggested to have great potential in this area. Thus, algae, filamentous fungi, yeast and bacteria have all received considerable attention and there is little question but that many of these can be used to make major contributions to the world protein pool. All have a common disadvantage, however, in the indigestibility or only slight digestibility of the rigid cell wall. Like protozoans, plasmodia, having no cell walls, would not have this disadvantage. Genin and Shepelev[107] have mass produced protozoa which were fed to men as a sole source of food and proved to be satisfactory, but aside from this Russian work, there seems to be little work designed to explore the possibility of using protozoans as food. Myxomycete plasmodia would have a distinct advantage over protozoans, since because of their larger size, harvesting problems would be much simpler.

As conducted by Brewer[37] and his associates, large-scale production of plasmodia would not be feasible at this time. Their medium contained large amounts of such relatively expensive components as casein hydrolysate, Difco yeast extract and recrystallized hemin converted to hematin and hence would make production prohibitive for economic reasons. However, there is no valid reason for believing that costs could not be markedly reduced. It seems possible that a cheap, crude source of hematin (or one of the substances which can be substituted for it) could be found. An encouraging fact is that several materials which can be substituted for methionine in the growth medium have already been found (Daniel and Babcock;[68] Henney and Lynch[140]). Because of their animal affinities, myxomycete plasmodia contain proteins which conceivably might be more suitable for use by man than many of the vegetable proteins which he presently uses. At this time the production of myxomycete plasmodia on a vast scale is purely speculative, but in view of their great potential and the present world protein shortage, further investigation of the large-scale production of these organisms might prove to be extremely valuable.

Before leaving the subject of the possibility of growing fungus mycelia or myxomycete plasmodia as a source of edible protein, attention must be called to the work of Fitzpatrick et al.[96] These investigators conducted feeding experiments with mushrooms *(Agaricus bisporus),* conducted amino acid analyses on the protein, and prepared purified protein of this fungus. They reported that the purified protein contained 11.79% nitrogen. This latter report needs verification, and if verified can raise some interesting points. It may simply mean that *Agaricus bisporus* protein is exceptional but it may mean that fungus proteins in general contain lesser quantities of nitrogen. Most reported crude protein values for fungi are based upon the assumption that these proteins contain about 16% nitrogen and hence crude protein values are calculated by determining Kjeldahl nitrogen and multiplying by the factor 6.25. If fungus proteins do in fact contain a lesser amount of nitrogen then all reported crude protein values are too low. If 11.79% is the correct quantity of nitrogen in fungus protein, then an accurate calculation for crude fungus protein would be made by multiplying Kjeldahl N by 8.48. Or, reported crude protein values (based on Kjeldahl N x 6.25) would have to be corrected by multiplying by the factor 1.356.

TOXIC SUBSTANCES IN FUNGI

With the great flood of research reports which has appeared in recent years on the subject of toxic substances from molds, mycotoxins, it would be impossible to prepare a review such as the present one without making some mention of these substances. Even if no such toxins had ever been reported, the writer would feel obligated to predict that some molds would be found that would have the capability of producing compounds toxic to other organisms. When one views the remarkably wide range of synthetic capacities of fungi of the mold type it seems almost inevitable that some poisonous compounds would be formed. Furthermore, in almost any related group of species there are usually a few to be found that have undesirable characteristics. Fortunately, however, in each group the number of undesirable species usually constitutes a quite small segment of the total number of species. For example, in the midwestern United States, the nature-lover can touch hundreds of different species of green plants with impunity; he gets poisoned only if he touches poison ivy. The wise nature-lover doesn't stop his nature-loving; he merely learns to recognize poison ivy and avoid it. Above all, he does not condemn all flowering plants because one happens to have the capability of causing him considerable discomfort.

A similar case in point is provided by the snake species native to the United States. Of the total number of different species a very small percentage is poisonous. An example of even greater pertinence in the present discussion is provided by those species we know as mushrooms. Of several thousand species of fleshy fungi usually called mushrooms, the deadly poisonous ones can be counted on the fingers of one hand, and the total number of poisonous ones (even to a slight degree) is probably less than a hundred.

What the reviewer is saying is precisely what he has said before (Gray[114]) and that is that a complete and proper assessment of the activities of the fungi as a whole reveals that insofar as man is concerned, the beneficial activities of fungi far outweigh their harmful activities. Unfortunately, the layman is usually unaware of this and after reading a newspaper or magazine article on poisonous molds, he frequently comes to the conclusion that all molds are poisonous and had best be avoided. Incorrect reporting does nothing to alleviate this common misbelief. For example, a national news magazine recently contained an article in which it was stated that Japanese foods are prepared with *Aspergillus flavus,* a mold which produces a toxin!

It is the contention of the present writer that to date the facts concerning the number of mold type fungi which have been proved to synthesize substances toxic to man and/or animals represents such a small percentage of the total number of species that to categorize molds in general as poisonous is unwarranted. In *World Protein Resources,* Wogan,[341] in a 20-page article devoted to mycotoxins in foodstuffs, lists only 11 mold species (in 6 genera) which have been definitely implicated in the production of toxic substances. Curiously enough, in the same volume in the paper immediately preceding Wogan's article, Liener[188] described 12 types of toxic and antinutritional substances synthesized in thirty-one commonly eaten (many by man) plants. These toxins include such substances as goiterogens, cyanogens, hemagglutinins, etc., and the plants listed included kidney beans, navy beans, garden peas, black-eyed peas, lima beans, peanuts, etc. *Surely nothing that grows on this earth is safe to eat!*

Admittedly there are molds that produce toxins. However, this writer's contention is that it is illogical, unscientific and possibly even extremely short-sighted to condemn all molds *per se* because some are known to synthesize toxins. It would seem more reasonable to first find out which molds produce toxins and then avoid them as one avoids poison ivy. Another point to consider is that some toxins may be produced only on particular substrates. For example, Challenger et al.[45] found that when *Penicillium brevicaule* is cultured on potato mash containing arsenious oxide, a toxic gas, trimethylarsine, is formed. Thus, it may be more than coincidence that aflatoxin was first discovered in peanuts on which molds had grown or that tempeh bongkrek may become poisonous while tempeh kedelee may not.

A variety of fungi has been reported to produce toxic substances but in a number of instances one is probably justified in questioning the identities of certain of the species that are reported to be toxin producers. A case in point is provided by gliotoxin. Weindling[329] reported that an imperfect fungus, *Trichoderma lignorum,* parasitized various pathogenic soil fungi by coiling hyphae around them or even killed them at a distance in submerged culture, which latter of course was evidence that a toxic substance was being produced. Weindling[330] later found the lethal principle to be formed by young hyphae only and he studied its decomposition in relation to oxygen, pH and heat. It was later found (Weindling[331,332]) that the organism involved was *Gliocladium fimbriatum* and not *Trichoderma lignorum,* and the name gliotoxin was applied to the toxic substance. It seems altogether possible that other fungi have been implicated as toxin producers in the same manner as *Trichoderma lignorum* and that certain species, now suspect, are not toxin producers at all. Johnson and Buchanan[163] reported the structure of gliotoxin to be as shown in A, Figure 39, but Bell et al.[22] later proposed that the structure was as in B. Johnson et al.[164] found that gliotoxin was toxic to rats and mice, but Larin (*c.f.* Taylor[301]) found that primates could tolerate doses of 0.2 mg/kg administered intramuscularly for 15 consecutive days. Insofar as this writer has been able to determine, there have been no reported outbreaks of animal poisonings due to gliotoxin nor have there been reported cases of humans being poisoned.

The term "mycotoxicoses" has been applied (Forgacs and Carll[101]) to the toxicity syn-

FIGURE 39

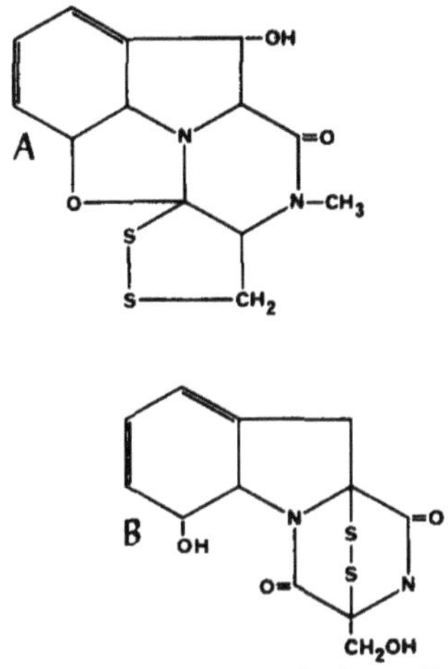

Structures proposed for gliotoxin, a metabolite of *Gliocladium fimbriatum*. A—according to Johnson and Buchanan;[163] B—according to Bell et al.[22]

dromes which result from ingesting foodstuffs which have been inadvertently contaminated with certain molds. Although the causal agent of ergot of rye, *Claviceps purpurea,* is not classified by the mycologist as a mold type fungus, ergotism in humans fits this definition of a mycotoxicosis and is briefly mentioned here because historically it represents our first documented example of poisoning due to fungus contamination of foodstuffs. *Claviceps purpurea* parasitizes rye and other grains, and as the black fungal sclerotia grow they replace the infected ovularies and thus may be harvested with the sound grains. If sclerotia are milled into flour with the harvested grain and a person ingests such contaminated flour for a period of time, the symptoms of gangrenous ergotism appear and may result in loss of limbs or death. According to Wogan[341] epidemics of ergotism were frequent in the Middle Ages, but the first association of the ergot fungus with ergotism was in the mid-sixteenth century. In spite of our long standing knowledge of the relationship between *Claviceps purpurea* and ergotism, it was not until the 1930's (Rothlin and Bircher[246]) that a group of six lysergic acid derivatives which are responsible for ergotism were isolated. According to Barger[17] the last major outbreak of ergotism in the United States was in 1825, but serious outbreaks occurred in Russia in 1926-27, in England in 1928, and an epidemic occurred in a village in France as recently as 1951 (Alexopoulos[1]). As a hazard to humans, ergotism is now virtually non-existent, but Garner[106] states that ergotism may still cause problems in veterinary practice.

In recent years the major emphasis in the area of fungus-produced toxic materials has largely been concentrated on toxins produced by mold types of fungi. Much of this work has been directly attributable to investigations stemming from livestock feeding on mold-contaminated grains or other feed seeds, hay or fodder. The vast majority of recent papers concerned with mycotoxicoses deal with a specific toxin, aflatoxin. However, a variety of other such toxicoses and fungus toxins has been reported and deserves some mention here.

Stachybotryotoxicosis

Stachybotryotoxicosis is primarily a disease of horses, the first case being reported in the Ukraine in 1931 (Vertinskii[316]). Apparently the disease is caused by a toxin which is produced by *Stachybotrys atra* when it grows on fodder, hay, etc. (Drobotko,[82] Drobotko et al.[83]). Forgacs et al.[102] tested a large number of strains of this species of imperfect fungus and found only a few to be toxic. Cattle, swine, sheep and other animals may be affected by the toxin (Forgacs et al.[102]). Fortuskny et al.[104] and Drobotko[82] reported that people exposed to aerosols from toxic substances (infected hay and straw) may be affected. When horses eat small amounts of infected material at intervals they usually die a slow death but when large amounts are ingested at one time, symptoms appear in 6 to 72 hours and death may occur in 17 hours (Clare[63]). This toxicosis is characterized by hemorrhage and necrosis of many tissues. Forgacs[100] has discussed the pathology and symptoms of stachybotryotoxicosis in some detail and for symptoms marking the progress of this disease as well as other details, the reader is referred to his paper.

Aspergillustoxicosis

From time to time there are outbreaks of what appears to be poisoning in livestock, and such outbreaks are often attributed to the ingestion of moldy feed. Thus, Carll et al.[42]

and Forgacs et al.[103] reported the isolation of strains of *Aspergillus chevalieri* and *Aspergillus clavatus* which when cultured on bread or grain and fed to calves produced clinical manifestations that resembled many of those of cattle which had eaten toxic foodstuffs in the field. In view of what is now known about aflatoxin, it seems possible that earlier suspected cases of poisoning may have been the result of ingesting feed contaminated with this toxin.

Moldy Corn Toxicosis

In late summer of 1952 an unknown disease of swine appeared in western Florida and was studied by Sippel et al.[269] This same disease was soon observed in Georgia. Apparently all of the animals had eaten moldy corn in the field. Sippel[268] concluded that the disease had first occurred as early as 1949 but was not diagnosed at that time. Burnside et al.[40] isolated 13 fungi from toxic field corn and found one strain of *Aspergillus flavus* and one of *Penicillium rubrum* which when cultured on sterilized corn and fed to swine produced toxicosis. Other outbreaks have occurred but, with the exception of one in 1955, this toxicosis has not been severe. However, Forgacs[100] states that other outbreaks may be expected to occur whenever climatic conditions become optimum for fungal growth and toxin formation. Young pigs are especially susceptible but older animals are also affected. The disease may be manifest in both acute and chronic forms. Autopsies on dead pigs revealed hemorrhages in many tissues. Forgacs[99,100] has presented detailed accounts of the pathology of moldy corn toxicosis and states that in many cases death is caused by hemorrhage. Diener and Davis[76] are of the opinion that aflatoxin may have been responsible for this type of poisoning.

Facial Eczema in Ruminants

This disease (incorrectly termed an eczema), while principally a disease of sheep, may also affect cattle and has been known in New Zealand since the beginning of this century. Apparently swine and horses are not susceptible. Dodd[80] and Taylor[301] have presented reports on the pathology and control of the disease as well as the chemistry of the toxin involved, and the present brief discussion is based primarily on their reports. The responsible fungus is *Pithomyces chartarum* (formerly known as *Sporodesmium bakeri*) which has a worldwide distribution and usually grows on any dead plant material. However, outbreaks of facial eczema usually occur only on rye grass pastures. Ingestion of toxic material leads to scab formation and exudation of fluid from skin that is exposed to sunlight (due to photosensitization). Liver damage also occurs and leads to occlusion of ducts and jaundice. Friedrichsons and Mathieson (*c.f.* Taylor[301]) have described the toxic substance (sporodesmin) as having the structure shown in Figure 40. Many severely affected sheep recover, but the only satisfactory control measure now known is to take sheep off pastures that are known or suspected to be dangerous.

Alimentary Toxic Aleukia (ATA)

Apparently this disease is the only one of the various mycotoxicoses which is known to have involved humans to any major extent. The disease seems especially prevalent in Russia and appears to be due to eating grains that have overwintered in the field. According to Joffee[162] this disease has occurred in Russia probably since the beginning of the nineteenth century but occurred with especial severity between 1942 and 1947, the peak year being 1944. In this latter year casualties greater than 1000 per 10,000 population are reported to have occurred in certain areas of the Orenburg district. Mayer[207] has reviewed 239 research papers on ATA, and the symptoms have been described by Forgacs[99] and Forgacs and Carll.[101] Primary pathologic changes are seen in blood-forming tissue.

Sarkisov[25] attributed ATA to a toxin pro-

FIGURE 40

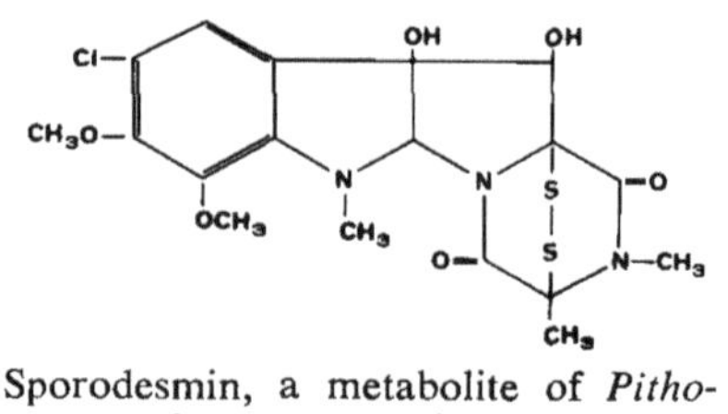

Sporodesmin, a metabolite of *Pithomyces chartarum*, which is involved in the syndrome known as facial eczema in ruminants.

FIGURE 41

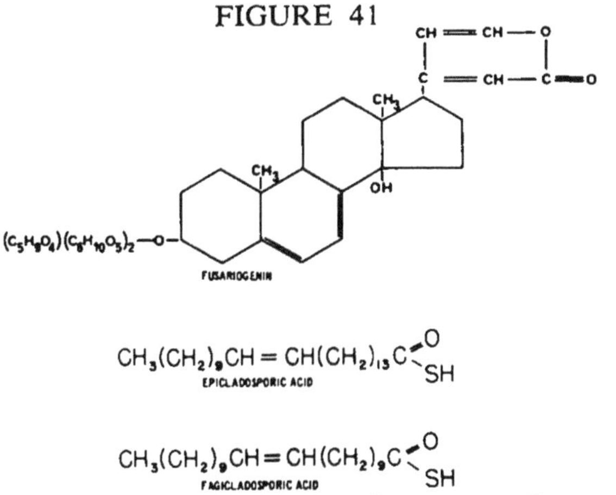

Toxic metabolites of fungi implicated by Joffee [162] in alimentary toxic aleukia (ATA).

FIGURE 42

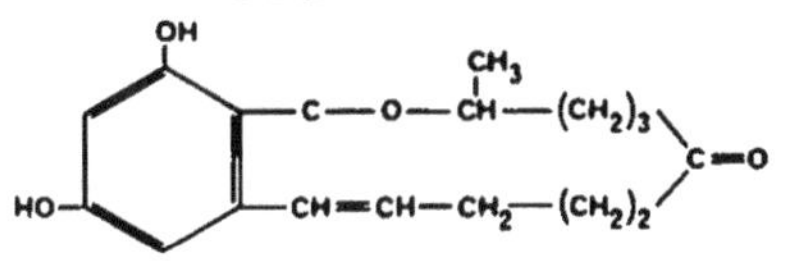

Metabolite of *Fusarium graminearum* involved in causing an estrogenic condition in swine.

duced by *Fusarium sporotrichioides* growing on grain, but Joffee[162] claims that many species of mold fungi may be involved and that the most important ones in causing ATA are species of *Fusarium, Cladosporium, Alternaria, Penicillium* and *Mucor. Fusarium poae, Fusarium sporotrichioides* and *Cladosporium epiphyllum* were cited as being especially toxic. Citing works of both Bekker[21] and Olifson,[223] Joffee[162] states that compositions of toxins produced by species of *Fusarium* and *Cladosporium* which he had isolated had been established (Figure 41). Joffee[161] studied the conditions for toxin formation and found it to be associated with the overwintering of grain in the field. Toxins were produced at temperatures below zero, but alternate freezing and thawing resulted in strongest toxin formation.

There is no doubting the reality and severity of ATA but its prevention is quite simply achieved: do not allow grain to overwinter in the field. The fungi which Joffee has implicated in ATA are termed "field" fungi by Christensen[59] and normally are not to be found on stored grain unless the moisture content is quite high. Therefore, if grain is harvested when mature and stored properly, ATA should present no special problem. The severe outbreaks in Russia during and after World War II were caused by a highly atypical set of circumstances: (1) allowing grain to overwinter in the field is not a usual agronomic practice and was probably necessitated by manpower shortages; (2) food was scarce in Russia and grain left in the field was probably used by many people; and (3) while the disease may affect anyone, undernourished persons are affected more severely.

Estrogenic Condition in Swine

McNutt et al.[209] observed that when swine consumed fungus-infested feed, sows developed enlarged vulvae, enlarged mammary glands, and in more severe cases a prolapse of the vagina and rectum occurred. Stob et al.[286] isolated a compound from corn infected with a *Fusarium* that apparently was the cause of the estrogenic syndrome. Christensen et al.[59] isolated an estrogen which they called "F-2" which is formed by isolates of *Fusarium graminearum* growing on autoclaved corn. Andrews[8] and Andrews and Stob,[9] Christensen et al.[59] and Urry et al. (*c.f.* Mirocha et al.[211]) have worked on the chemical identity of the estrogenic metabolite and in 1967 Mirocha et al. reported it to be one of the enantiomorphs of 6-(10-hydroxy-6-oxo-trans-l-undecynl)-β-resorcylic acid lactone (Figure 42).

Mirocha and his associates studied the estrogenic response of rats to this compound and found a direct linear relationship between estrogen concentration and uterus weight. However, with high concentrations there was no increase in body weight while low concentrations caused a relative increase in body weight. Thus, if this compound is used in feed as a growth promoter, care would have to be exercised. As in the prevention of ATA, the estrogenic condition in swine can be prevented by the simple expedient of making sure that they do not eat moldy corn.

Toxic Moldy Rice

Since Oriental peoples customarily eat a wide variety of fungus-fermented rice and soybean products, recent investigations of toxin-producing fungi have served to focus attention on the possibility that moldy rice may be the cause of human poisoning. Thus, Kinosita and Shikata[167] report that 12 species and 1 variety of *Aspergillus* and 12 species of *Penicillium* produce toxic metabolites, listing such well-known substances as patulin, kojic acid, citrinin, emodin, etc. Oddly enough, they list several species which Diener et al.[77] cultured on peanuts which they fed to Peking ducklings and obtained greater weight gains than they obtained with control rations.

Miyake and Saito[212] state that there are more than 15 fungi which cause moldy rice and regard *Penicillium citreo-viride, Penicillium citrinum* and *Penicillium islandicum* as having representative toxicities. They fed rice on which *Penicillium islandicum* had been grown to 30 rats, and 29 rats died between the 20th and 736th day. Of the rats that died, two were proved to have typical hepatomas and three were proved to have sarcoma-like tumors. Two toxic metabolites, luteoskyrin and a chlorine-containing peptide (Figure 43), have been isolated from *Penicillium islandicum* (Kobayashi et al.,[173–175] Uraguchi et al.,[309,310] Marumo,[204] Shibata and Kitagawa[260,261]). Miyake and Saito list an LD_{50} for luteoskyrin of 1.47 subcutaneously and one of 2.21 mg/10g orally for rats and mice, and LD_{50} values of 4.75 and 65.5μg/10g for the chlorine-containing peptide (islandicum). Because of the higher incidence of liver cancer in the rice-eating Orient, Kraybill and Shimkin[178] believe that there may be some relationship between liver injury and high concentrations of moldy diets.

On the other hand, Yokotsuka et al.[348] tested 73 strains of Japanese industrial molds used in the production of shoyu, miso and alcoholic beverages and none of them produced aflatoxins, although some did produce fluorescent compounds. These investigators admitted the possibility that koji cultures might conceivably become contaminated occasionally with aflatoxin-producing mold strains, but they implied that the possibility was a remote one in view of the enormous numbers of spores of non-aflatoxin-producing strains used in the production of seed cultures. Several other investigators have confirmed the non-toxic nature of these industrial molds. Since many Oriental fungus-fermented foods are prepared with an *Aspergillus,* some attention has been given also to the possibility that aflatoxin may be in such foods. However, this toxin has not been detected in any of the samples which were examined (Uritani[311]).

FIGURE 43

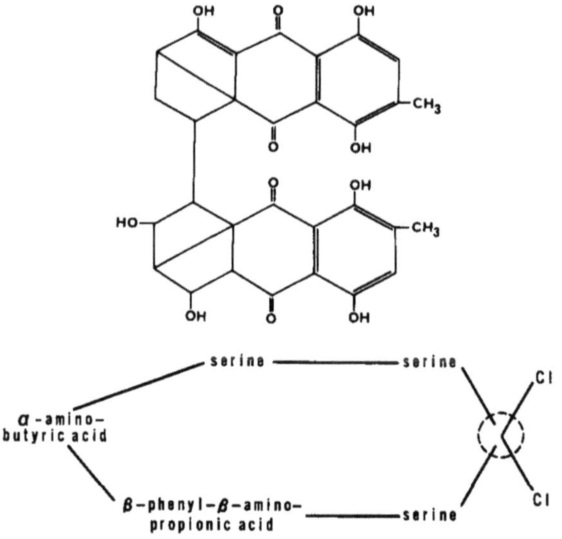

Toxic metabolites produced by *Penicillium islandicum.* Above, luteoskyrin; below islanditoxin, a chlorine-containing peptide.

Aflatoxin and Other Toxins

The fungus toxin which has received the greatest amount of attention from the greatest number of investigators is aflatoxin. More than any other event the discovery of this compound has served to focus attention on the fact that some fungi are poisonous. Since Blount's[32] first report in 1961, such an astonishingly great number of workers (a recent book on aflatoxin edited by Goldblatt[110] lists 995 names in the author index) have flung themselves into aflatoxin investigations that it would almost appear that we are witnessing a trend in science. However, having witnessed similar efforts in connection with such subjects as krillium, gibberellin, hydroponics, etc., this reviewer is more inclined to believe that we are witnessing a fashion in science and that, ultimately, aflatoxin like the others will be placed in proper perspective. Unless specific evidence is soon provided linking aflatoxin causally with human liver cancer, it seems likely that we will be hearing less and less about this compound. Unfortunately, this great emphasis on a few mycotoxins and especially aflatoxin has had its impact on attempts by various investigators to point out the great potential resident in many fungi in the area of protein synthesis.

The literature on aflatoxin and other mycotoxins is now voluminous. For example, Wogan[340] edited a book entitled *Myxotoxins in Foodstuffs* in which there are 22 contributed papers with a total of 349 literature citations. Two years later, Mateles and Wogan[205] edited another volume: *Biochemistry of Some Food-borne Microbial Toxins,* the fungal toxins section (pp. 69-168) of which consisted of papers by six contributors who in their collective bibliographies cited 333 references. The index of the first book listed 29 species of fungi, that of the second book only 16. In the three years since the appearance of the latter of these two books, many additional papers and another book (Goldblatt[110]) have appeared, so it is obvious that complete coverage of extant literature cannot be made here. On the contrary, an attempt will be made to present a brief history of the discovery of aflatoxin (the most publicized of the mycotoxins) as well as other such compounds, to enumerate some of the fungi which appear to be implicated, to point out that aflatoxin is carcinogenic in some animals, but, above all, to make a plea for a generally saner view of the fungi, since at this point it would appear that moldy Brazilian peanut meal may have caused an unwarranted degree of panic insofar as man's health and well-being are concerned.

There is no question but that certain fungi produce toxic compounds and that aflatoxin has caused serious economic loss. Attention was first called to aflatoxin when it caused the deaths of about 100,000 turkey poults in England (Blount,[32] Sargeant et al.,[250] Lancaster et al.,[186] Allcroft and Carnaghan,[3] etc.). The initial report of the death of turkeys was followed in short order by a report of the death of 14,000 ducklings on one farm (Asplin and Carnaghan[14]) and nine outbreaks of disease in calves (Loosmore and Markson[197]). In all instances these events were associated with the feeding of Brazilian peanut meal. Sargeant et al.[250] isolated a crystalline substance from peanut meal and found that 20μg administered orally was fatal to day-old ducklings in 24 hours. The material was suspected to be a fungal metabolite, and a common storage fungus, *Aspergillus flavus,* was isolated and identified. The toxic material(s) was named aflatoxin and typical of its effects were liver lesions which indicated that it was carcinogenic. Deaths of turkeys, pigs and calves (Allcroft and Carnaghan[3]) were attributed to aflatoxin, but chickens were reported to be comparatively resistant to the toxin, although those which had been fed 15% toxic peanut meal showed definite hepatic changes. Cottier et al.[66] found that mortality of chickens fed at a level of 308 p.p.b. aflatoxin showed only a slightly higher mortality than those on control rations. However, mortalities were higher in groups fed rations containing 610 and 1834 p.p.b. Kraybill and Shimkin[178] listed the general order of susceptibility of poultry to aflatoxin as: duckling $>$ turkey $>$ chicken and for larger animals: swine $>$ cattle $>$ horses $>$ sheep. At about the same time that Turkey "X" Disease appeared in England, there was a widespread outbreak of trout hepatoma in the United States. Wolf and Jackson[343] showed that the causal agent was in the cottonseed meal of the ration, and Jackson et al.[156] and Sinnhuber et al.[267] demonstrated the presence of aflatoxin B_1 in the cottonseed meal. The toxicity and carcinogenicity of aflatoxin have since been demonstrated with a variety of experimental animals by a number of investigators.

According to Wilson et al.[339] there are at least eight closely related difuranocoumarin compounds known as aflatoxins, the four most familiar ones being aflatoxins B_1, B_2, G_1 and G_2. The structures of these four compounds have been established (Figure 44) through the efforts of Asao et al.,[11-13] Chang et al.,[46] Cheung and Sim,[57] Hartley et al.,[135] Nesbitt et al.[218] and Van Dorp et al.[314] The works of Holker and Underwood,[146] Biollaz et al.[23,24] and Donkersloot et al.[81] leave little doubt that aflatoxin is synthesized by way of acetate. This latter fact may be of some significance, since the growth-retarding effect of acetate on fungi

FIGURE 44

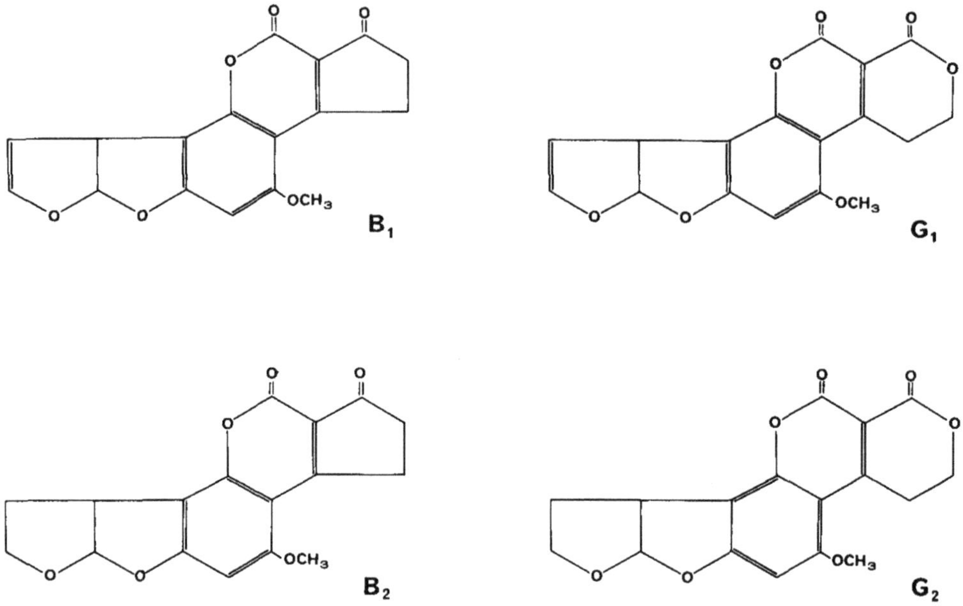

Structures of the common aflatoxins.

has long been known (Kirby et al.[168,169]).

While the total number of mold type fungi variously reported to produce toxins seems large (at least 41), in terms of the total number of fungus species it is extremely small. If the conservative estimate of Martin[202] is used, the reported toxic species represent about 0.05% of the total. In the index to Mateles and Wogan's[205] volume, only 16 mold species are listed and of these only 6 are noted to be toxic. Wogan's[340] index lists 29 species (five in common with Mateles and Wogan) and of these, 22 are reported to produce toxins. Curiously enough, only one of the species listed by Joffee[162] is listed in the index to Wogan's volume although Joffee's paper appears in this volume. Joffee lists 27 species which he reported to be toxic on the basis of results obtained when he fed culture filtrates to rabbits and guinea pigs. Unfortunately, Joffee does not present exact details as to his culture methods and, hence, it is impossible to judge whether or not toxic compounds may have been produced primarily as a result of the culture methods used. In Table 15 are listed all species noted by Joffee as well as those appearing in the indices to Wogan and Mateles and Wogan.

Apparently there is some disagreement among different investigators concerning which fungi produce toxins. For example, in Table 15 ten species of *Penicillium* and six species of *Aspergillus* are listed as toxin producers. However, Parrish et al.[226] screened 166 strains representing 14 species of *Aspergillus* and 8 species of *Penicillium* and found that aflatoxins were produced only by *Aspergillus flavus* and *Aspergillus parasiticus* (these are probably synonyms). Furthermore, the mere fact that a fungus is *Aspergillus flavus* does not necessarily mean that it will produce aflatoxin since only 26 (28.3%) of 93 strains of *Aspergillus flavus* produced aflatoxin. Similarly, Wilson et al.[339] studied aflatoxin production by 121 isolates representing 29 species in 7 genera (mostly *Aspergillus* and *Penicillium*) and demonstrated this property only in *Aspergillus flavus* and *Aspergillus parasiticus,* surely a small harvest for the toxin hunter! Such reports as these do not necessarily negate the reported toxicities of all the species listed in Table 15, since it is possible that many of the fungi listed there may have been producing mycotoxins other than aflatoxin.

Since the real criterion of toxicity must ultimately rest upon feeding trials, and since aflatoxin production appears to be frequently associated with moldy peanuts, the work of

TABLE 15

Fungal Species Listed

Fungal species		Fungal species	
Alternaria tenuis	+	*Mucor hiemalis*	+
Aspergillus candidus	−	*Mucor racemosus*	+
Aspergillus chevalieri	+	*Nectria radicicola*	−
Aspergillus clavatus	+	*Neurospora sitophila*	−
Aspergillus flavus	+	*Paecilomyces varioti*	+
Aspergillus fumigatus	+	*Penicillium brevicompactum*	+
Aspergillus glaucus	−	*Penicillium chartarum*	+
Aspergillus halophilicus	−	*Penicillium citrinum*	+
Aspergillus niger	−	*Penicillium crustosum*	+
Aspergillus ochraceus	+	*Penicillium expansum*	−
Aspergillus oryzae	−	*Penicillium islandicum*	+
Aspergillus parasiticus	+	*Penicillium jenseni*	+
Aspergillus restrictus	−	*Penicillium notatum*	+
Aspergillus soyae	−	*Penicillium patulum*	+
Chaetomium cochilodes	−	*Penicillium rubrum*	+
Cladosporium epiphyllum	+	*Penicillium steckii*	+
Cladosporium fagi	+	*Penicillium terlikowskii*	−
Cladosporium fuligineum	+	*Penicillium toxicarum*	+
Cladosporium gracile	+	*Piptocephalis fresiana*	+
Fusarium equiseti	+	*Pithomyces chartarum*	+
Fusarium graminearum	+	*Rhizopus nigricans*	+
Fusarium lateritium	+	*Rhizopus oligosporus*	+
Fusarium nivale	+	*Rhizopus oryzae*	−
Fusarium poae	+	*Sclerotinia sclerotiorum*	−
Fusarium roseum	+	*Sporodesmium bakeri*	+
Fusarium sambucinum	+	*Stachybotrys atra*	+
Fusarium semitectum	+	*Thamnidium elegans*	+
Fusarium sporotrichioides	+	*Trichoderma lignorum*	+
Fusarium trincinctum	+	*Trichoderma viride*	+
Gibberella zeae	+	*Trichothecium roseum*	+
Mucor albo-ater	+	*Verticillium lateritium*	+
Mucor corticola	+		

Wogan,[340] Joffee [162] and Mateles and Wogan [205]

+ indicates that the species was reported to produce a toxin, − indicates that no such report was found

Diener et al.[77] is of especial interest. These workers cultured eight species of *Aspergillus* and one of *Penicillium* (all of which had been isolated from peanuts) on autoclaved peanuts. After three weeks the molded peanuts were oven-dried at 135°C and were then defatted by extraction with petroleum ether. These materials were then used (in the amount of 40% by weight) to compound rations for ducklings. Two-day old Peking ducklings were fed on commercial "starter" for two days and then on a ration containing moldy peanuts for three days after which time they were weighed. Results are presented in Table 16.

From the data presented in Table 16 it is evident that only the feeding of *Aspergillus flavus* resulted in the death of ducklings. Furthermore, during the course of the short feeding trial, six of the nine fungi used as food produced greater weight gains than control rations containing no fungus. It is also interesting to note that ducklings fed on peanuts on which *Aspergillus restrictus* had grown showed a 25% higher weight gain than those fed the control diet. Of even greater interest is the finding that ducklings showed less weight gain on the control diet than on the one containing *Penicillium citrinum*, a species reported by

TABLE 16

Weight Gains of Peking Ducklings Fed on Rations Containing Moldy Peanuts

Rations molded by	Av initial wt/bird (g)	Av wt gain/bird (g)	Wt gain as percentage of control
Aspergillus amstelodami	79	64	106.6**
*Aspergillus flavus**	79	−13	——
Aspergillus chevalieri	79	58	96.6
Aspergilus ruber	79	53	88.3
Aspergillus terreus	79	70	116.6**
Aspergillus restrictus	80	75	125.0**
Aspergillus candidus	80	61	101.6**
Aspergillus micro-virido-citreus	80	63	105.0**
Penicillium citrinum	81	71	118.3**
Control (no fungus)	80	60	100.0

Based on Diener et al.[77]

*All birds died
**Greater weight gain than controls

Majumder et al.[200] to produce toxic metabolites when grown on rice. From these reports it can only be inferred that ducklings are resistant to whatever toxin is produced, that the strain used by Diener and his associates was different from that used by Majumder et al., or else that the metabolism of *Penicillium citrinum* changes with substrate.

No phase of fungus metabolism is more susceptible to alteration due to slight changes in environment than carbon metabolism, and it is entirely possible that in many instances mycotoxins are synthesized as a result of a unique set of environmental conditions. This is especially true when a fungus is cultured under conditions of high carbohydrate concentration. With very few exceptions the fungi which are present on the earth today are those capable of efficiently converting carbon from a very dilute substrate to tissue carbon. In general, if other conditions are favorable, at low carbohydrate concentration the principal products of fungus growth are mycelium and respiratory carbon dioxide. As carbohydrate concentration is increased, the Economic Coefficient value also increases; i.e., the organism converts a lesser percentage of substrate carbon utilized to mycelium carbon and in many cases a greater percentage to other metabolites which are not assimilation products. Foster[105] claimed that under "luxury" conditions of carbohydrate concentration, carbon metabolism becomes deranged, a view which is borne out by the findings of Davis et al.[73] These workers investigated the relation of sugar concentration of the medium to aflatoxin production, and the results are presented graphically in Figure 45. From this figure it may be seen that aflatoxin production increased in the range of sucrose concentration of 0 to 20, but large amounts were synthesized even at sucrose concentrations as high as 50%. On the other hand, much smaller amounts were produced in medium of lower sucrose concentration, and there is a strong possibility that even lesser amounts might have been produced had their cultures been agitated rather than still.

Diener and Davis[76] have recently published an extensive review of aflatoxin formation and suggest that for the production of large quantities of aflatoxins the following points should be observed: (1) use of rice or shredded wheat as substrate; (2) use of *Aspergillus flavus* NRRL 2999, NRRL 3145 or NRRL A-13794; (3) maintenance of high substrate moisture; (4) maintenance of an atmosphere of 99-100% relative humidity; and (5) incubation at 28-30°C for one week. Forgacs[99] states that even for those fungi which produce toxins, there appear to be specific conditions neces-

FIGURE 45

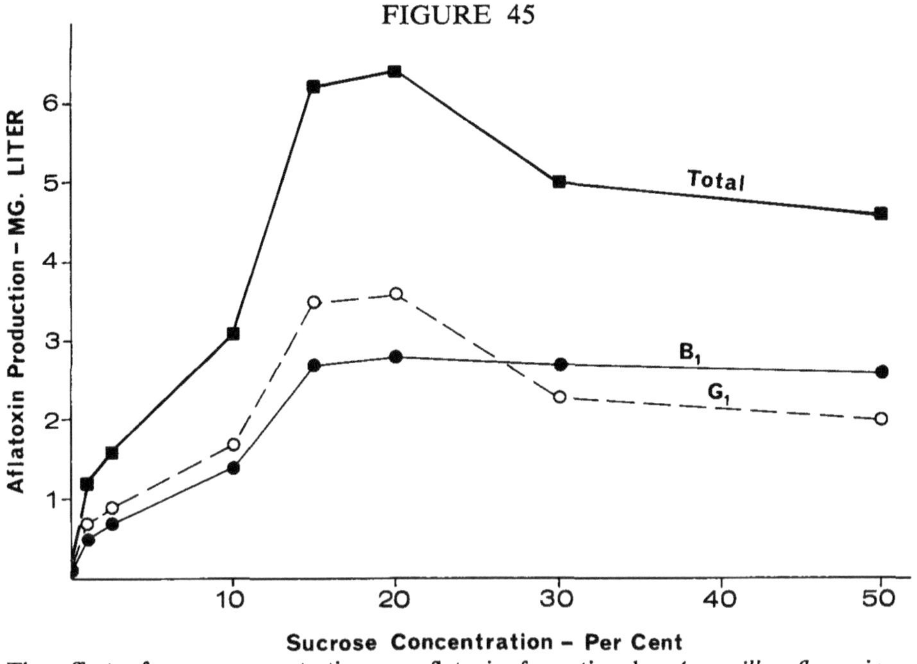

The effect of sugar concentration on aflatoxin formation by *Aspergillus flavus* in a semisynthetic medium. Based on data of Davis et al.[73]

sary for toxin formation. The fact that Diener and Davis recommend the use of only two substrates is reminiscent of the findings of Iizuka and Iida.[155] Working with *Aspergillus oryzae* var. *microsporus* these latter workers found that it produced a strong toxin only when it was cultured in media containing malt extract. The structure of the toxin was established and it was named maltoryzine. Schroeder and Hein[253] suggest that aflatoxin synthesis is a function of growth rate and that synthesis increases when metabolic activities are accelerated by increasing temperatures. They further suggested that aflatoxins are primarily a problem of tropical and subtropical regions. It would seem that at this time a most profitable line of investigation would be to attempt to determine the specific field and storage conditions under which aflatoxin may be synthesized.

Since *Aspergillus flavus* is an ubiquitous fungus, the prediction of Hesseltine et al.[144] that aflatoxin would eventually be found in many agricultural commodities seemed a very reasonable one and has since been proved to be true. In Table 17 are listed various materials in which aflatoxin has been found. In addition to the commodities listed in this table, Allcroft and Carnaghan (*c.f.* Kraybill and Shimkin[178]) have reported the occurrence of aflatoxin in milk from cows fed on toxic peanut meal, and Allcroft et al.[4] have reported its occurrence in sheep's milk. The evidence now at hand clearly indicates that aflatoxin is an extremely dangerous carcinogenic substance insofar as a number of animals are concerned, and, through the poisoning of poultry and larger animals, has been responsible for considerable economic loss. However, there is no evidence that man has been affected by this mycotoxin. The full impact of the early report on aflatoxin will, of course, never be known. The value of peanut flour in the diets of malnourished children was well known, but agencies such as WHO, FAO and UNICEF (Milner[210]) agreed that pending further clarification of the aflatoxin situation they would not press various programs involving the use of peanut flour for infant feeding. Under the circumstances there seemed to be no reasonable course to take other than this one, and this action may have prevented the poisoning of many children in underdeveloped countries or it may have for a time prevented them from getting the much-needed protein which can be obtained from peanuts. In all probability the latter situation obtained, since it seems inconceivable that any reliable agency would have assumed the responsibility for feeding accidentally molded food to children, no matter how poor their lot may have been. It would now appear that a considerable amount of panic was involved in the situation and, to paraphrase Shakespeare, there may have been "Much ado about peanutting!" To date there seems to be little or no evidence that aflatoxin constitutes any problem insofar as humans are concerned. Differences in susceptibility of various animals have already been noted, and Allcroft and Carnaghan[3] state, "There is no evidence that man is susceptible to the toxin . . .," and in the same year Spensley[279] reported, "There is no evidence that human beings have suffered. . . . No sample of edible peanuts in Britain has yet been found to be toxic." Kraybill and Shimkin[178] state that, "No information is available regarding mycotoxicoses in man caused by contaminated peanuts or other cereal products, or products made from these materials." Wogan[340] refers, without explanation, to one instance of mycotoxicosis in man, but on November 12, 1969, (personal communication) stated: "In response to your question, I know of no new documented cases of human poisoning attributable to mycotoxins." Thus, there is still no real evidence that aflatoxin did or does now constitute a problem to man. The fact that the incidence of liver cancer is higher in Africa and in the rice-eating Orient, where moldy foods are more commonly eaten, does suggest that there may be some relationship between liver cancer and mycotoxins; however, such a relationship is far from being proved. Protein malnutrition is widespread both in Africa and in the Orient, and it is possible that mycotoxins may affect man under such conditions. Madhavan and Gopalan[198] compared the effect of aflatoxin on rats fed a low protein diet with the effect on those fed a high protein diet and found that there was greater liver damage in the protein-deficient animals. Similarly, Madhavan et al.[199] have shown that reduced protein intake increases the susceptibility of rhesus monkeys to aflatoxin. When administered aflatoxin in

TABLE 17

Some Foods and Feeds in Which Aflatoxins Have Been Found

Food or Feed	References
barley	Wogan[342]
cassava	Borker et al.;[35] Wogan[342]
corn	Borker et al.;[35] Wogan[342]
coastal bermudagrass hay	Diener and Davis[76]
cottonseed	Wogan[342]
cottonseed meal	Sinnhuber et al.;[267] Borker et al.;[35] Jackson et al.;[156] Diener and Davis[7]
cowpeas	Wogan[342]
grain sorghum	Shotwell et al.[262]
millet	Wogan[342]
oats	Diener and Davis[76]
peanut butter	Eadie and O'Rear[88]
peanut meal	Borker et al.[35]
peanuts	Allcroft;[2] Eadie and O'Rear;[88] Wogan;[342] Taber and Schroeder[297]
peas	Borker et al.;[35] Wogan[342]
rice	Borker et al.;[35] Wogan[342]
sesame	Wogan[342]
sorghum	Wogan[342]
soybean meal	Diener and Davis[76]
soybeans	Borker et al.;[35] Wogan[342]
spaghetti (dry)	Walbeek et al.[325]
sweet potatoes	Wogan[342]
wheat	Borker et al.[35]

the amount of 100 μg per day, animals on a low protein diet died while those on a high protein diet survived. If the same situation obtains in man, then the real problems seem to be that of providing adequate supplies of protein in those areas where it is in insufficient supply. Vigorous educational programs should be instituted in underdeveloped countries with the specific objective of pointing out the potential dangers associated with the eating of moldy or improperly handled food. Such a program would require time and much patience, since it should be borne in mind that the mental capacities of people suffering chronic protein deficiency are not as great as those of properly nourished people. Thus, again, we return to one of the major theses of this review and that is that perhaps the major problem facing the world today is that of supplying sufficient protein for the growing population.

One cannot but admire the fine efforts that have gone into the production of the hundreds of research reports that have contributed so much to our knowledge of the culture and metabolism of *Aspergillus flavus* and related species, the pathological effects of aflatoxin on animals, and the elucidation of the structures of the various aflatoxins as well as their mode of synthesis. Nonetheless, the fact still remains that at this time aflatoxin does not seem to present any problem to man *per se*. There is no question but that various animals are affected by this carcinogen; however, solution of that problem seems relatively simple, at least in theory. The old practice of feeding accidentally molded food to livestock (or people) could well stand re-examination and should promptly be discarded. Prompt harvesting, proper drying, and adequate storage procedures would all help prevent the growth of *Aspergillus flavus* on foods and feeds, and the only times when aflatoxin would be produced would be when environmental conditions were such that proper harvesting, drying and storage procedures could not be adhered to or when human ignorance or human greed entered the picture. As noted earlier because of the much higher incidence of liver cancer among Bantu Africans, it has been suggested that this situation may be related to the ingestion of mycotoxins. However, no one appears as yet to have had the desire or the temerity to suggest that the much lower incidence of leukemia and cancers of the stomach, large bowel, breast, endometrium, ovary and brain among Africans (Oettle[22]) may bear the same relationship to mycotoxins!

One additional mycotoxin deserves brief mention here since it seems to be gaining in attention although certainly not to the extent to which aflatoxin has. Aware of the work of Miyake and his associates in Japan, Isaacson (*c.f.* Purchase and Nel[233]) suggested that the high incidence of liver cancer in the Bantu population of South Africa might be related to mold metabolities. Scott[254] was able to show that 22 strains of fungi isolated from South African cereals and legume products were toxic. When corn meal was inoculated with one of the strains of *Aspergillus ochraceus,* a toxic metabolite called ochratoxin A was formed. This material was shown by Van der Merwe et al.[312] to have the structure shown in Figure 46. Two related compounds, ochratoxins B and C, were also isolated but these were less toxic than ochratoxin A. Van der Merwe et al.[313] reported LD_{50} values for ochratoxin A for day-old ducklings to be about 25μg, but Purchase and Nel[233] claim that this value is too low and that the true value is about 150μg per duckling. These latter workers also reported that the acute toxicity for 150g albino rats is about 20 mg/kg per os and that ochratoxin A causes tubular necrosis of the kidney, mild degeneration of the liver, and enteritis. In order to obtain large quantities of ochratoxin for studies of its toxicological properties, Ferreira[95] studied its production in relation to nitrogen source, glutamic acid concentration, lactic acid addition, carbon source and sucrose concentration. Highest yields reported were 100 mg per liter of medium in which sucrose and glutamate were the carbon and nitrogen

FIGURE 46

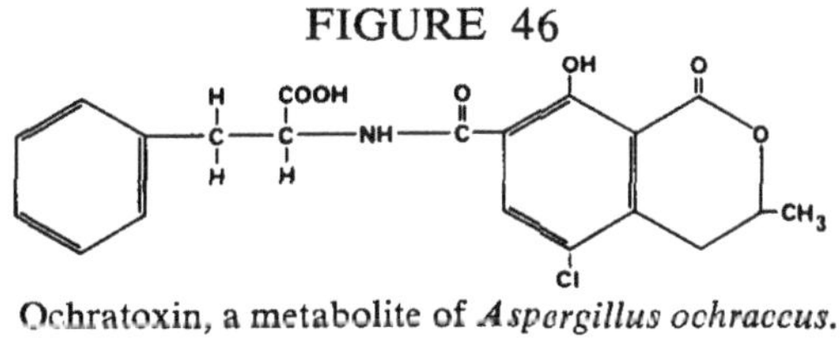

Ochratoxin, a metabolite of *Aspergillus ochraceus*.

sources. Davis et al.[74] were able to obtain higher yields in a semisynthetic medium, 29 mg per 100 ml of medium in still culture. In their studies of the biogenesis of ochratoxin A, Searcy et al.,[256] using labeled phenylalanine and acetate, found that phenylalanine could be incorporated unaltered into the phenylalanine moiety of ochratoxin but that the isocoumarin moiety was derived mostly through acetate condensation. Thus, we are obtaining more and more information concerning the metabolism of *Aspergillus ochraceus,* especially with regard to its synthesis of ochratoxin A. However, Purchase and Nel[233] state, ". . . we are a long way from showing any relationship between hepatoma and mycotoxins. . . ."

For the reader who is interested in more extensive discussions of mycotoxins as well as bacterial and algal toxins, the works of Wogan,[340] Mateles and Wogan[205] and Goldblatt[110] should be consulted.

SUMMATION

From the above account it is obvious that although filamentous fungi of various types have been widely used in nearly all areas of the world either directly as food or as a means of processing food, in terms of total food consumption by man, the contributions of these organisms have been relatively small. Thus, while the statistics which show us that in fiscal 1967-68 225 million pounds of mushrooms were consumed in the United States appear rather impressive, the contribution of these items to the overall food supply is minute. When it is realized that about 330 billion pounds of food are consumed annually by the citizens of this country, this seemingly great amount of mushrooms constitutes less than seven one-hundredths of one per cent of the total food consumed. Fungi have been little used as domestic animal food and their use in this manner constitutes a virtually unexplored area.

There is no question but that production of fleshy fungi for direct use as food for humans could and should be increased several-fold. This is especially true for Oriental countries where protein is already in short supply and much of the current edible mushroom production is conducted by primitive methods. A tenfold increase in production of both the shiitake and the padi straw mushroom in the Orient would contribute measurably to the protein supply in areas where protein malnutrition has been chronic for many years. That such expanded production could be accomplished readily is exemplified by the phenomenal development of the common cultivated mushroom industry which has occurred in Taiwan during the past decade. There seems to be little justification for attempting to expand the truffle industry, unless methods can be developed (and because of the nature of truffle production, this seems highly improbable) by which these delicious mushrooms can be produced as cheaply and in as great quantity as the other three commonly cultivated mushrooms. Some work has been conducted on the production of compost for mushroom culture from such waste materials as sawdust, garbage, old newspapers, etc. It is imperative that such work be continued and greatly expanded since it offers an excellent opportunity to demonstrate the old adage of killing two birds with one stone, in this instance, food production and waste disposal. However, in spite of the waste disposal and pollution problems now faced by virtually all municipalities, very few have been prepared educationally and psychologically to support developments of this type even though they appear to have great potential.

A sharp distinction exists between the Occident and the Orient with respect to the nature of the substrates which are processed with fungi for the production of various foodstuffs. In the Occident, milk proteins are processed with mold type fungi to form two basic cheese types, Roquefort and Camembert, while in the Orient vast quantities of soybeans, rice and wheat are processed with fungi in the preparation of a great variety of different foodstuffs. While some present day toxin hunters apparently have seen a toxic molecule under every fungal hypha and, hence, have succeeded in arousing suspicion concerning the many fungus-processed foods of the Orient, this reviewer has a completely different point of view. He prefers to think of the development of fungal methods for processing soybeans and/or rice for the purpose of producing a variety of different food items as a blessing which for centuries has helped prevent diet monotony in the Orient and, at the same time, has in many instances provided means of producing more easily digestible and more tasty foodstuffs. It would be expected that, occasionally, some food of this type, often crudely prepared by unlettered people, might become contaminated by an undesirable or even a toxin-producing microorganism. The obvious course of action (and this course is being followed by Hesseltine and his associates and others in this country) is to attempt to place the production of such foods on a more scientific and sanitary basis. However, before too many accusing fingers are pointed at the Indonesian peasant who prepares tempeh in a banana leaf, it should be remembered that not too many years have elapsed since a number of individuals died annually in the United States as a result of botulinus poisoning due to the ingestion of home-canned green beans. With a bit of reflection the older reader may also recall that occasional ptomaine poisoning was an unpleasant part of the American scene.

While much research yet remains to be done, methods are now available by which almost unlimited quantities of protein can be produced by culturing fungal mycelia submerged in liquid media with a cheap or waste carbohydrate as the carbon source and an inorganic nitrogen salt as the nitrogen source. We may find that in the not too distant future it may be necessary to run separate lines from kitchen garbage disposal units to fungal, yeast or bacterial processing plants where microbial proteins can be made and, at the same time, prevent sewage disposal plants from being overloaded. The greatest amount of research has been conducted with mycelia of fleshy Basidiomycetes and Fungi Imperfecti (both of which have advantages and disadvantages); however, there is no reason to believe that many Ascomycetes and Phycomycetes could not be used in similar manner. Such a process provides means of increasing the absolute amount of protein which can be produced per unit area of land and also provides means of producing protein from materials which are now allowed to go to waste. Whether or not such protein can be produced economically depends wholly upon one's point of view. In terms of cost per pound, it is improbable that at this stage fungal protein could be produced in many areas at a cost which would enable it to be competitive with soybean protein. When one takes a broader viewpoint and recognizes that most of the trouble spots in the world today are in areas where there is chronic protein malnutrition and that such trouble may in large measure be related to this protein insufficiency, then the production of microbial protein at twice the cost of soybean protein may seem cheap. One is justified in wondering if a Castro could have happened in a well-nourished Cuba, and it is most revealing to calculate the cost in dollars of doubling the per capita protein availability in Vietnam, North and South, and compare this figure with the $30 billion it costs the United States annually to maintain the unpopular war there.

Exactly when fungi as well as other microorganisms of potential will be exploited fully as sources of protein cannot be predicted. However, it can be stated with a reasonable degree of certainty that it will not occur until there is a much greater general awareness of the fact that the two greatest dangers facing the world today are not Communism and mycotoxins but rapidly rising population and protein malnutrition. Unfortunately, the awareness of reality is often influenced by political actions of governments which latter more frequently than not bear no resemblance to reality. In the meantime this reviewer unequivocally opts for the men and agencies who had the guts to fly real food, not committees, to starving Biafran children or, for that matter, any starving people regardless of the form of government under which they may live.

A considerably greater amount of space in the present review has been devoted to a discussion of mycotoxins than was originally intended. This has been done for two reasons: (1) to describe some of the very excellent work that has been conducted recently on fungal metabolism; and (2) to point out that, with the exception of toxic alimentary aleukia (ATA), there is very little convincing evidence to connect fungal toxins with various human aliments. This reviewer has witnessed over half a century of attempts to relate cancer causally with a variety of different agents and he is not prepared at this point to quickly accept the view that the higher incidence of liver cancer among Bantu Africans than among Afro-Americans is necessarily due to the ingestion of mycotoxins, especially if he has to accept the assumption suggested by Kraybill and Shapiro[177] that there has been no change in the genetic pool of the American Negro after several centuries of rather intimate biological contact with other races in the New World. However, there is no question but that several mycotoxins, and especially aflatoxin, affect a variety of animals. In most instances the accidental poisoning of livestock can probably be prevented if greater care is exercised in the feeding of livestock. Whether or not various mycotoxins can be demonstrated to have a causal relationship with liver cancer cannot be predicted; however, to date all evidence of such a relationship is circumstantial and in most instances quite nebulous.

At this time the fungi must be considered one of man's few remaining relatively unexplored food resources. How long they will so remain cannot be predicted.

Acknowledgments

The reviewer wishes to express his thanks for the assistance given him by colleagues all over the world in supplying information and certain of the photographs. He is especially indebted to Mr. Dan Irwin and Mr. Su-Hwa Lee for their assistance in preparation of many of the illustrations.

REFERENCES

1. Alexopoulos, C. J., *Introductory Mycology*, John Wiley & Sons, New York, 1952.
2. Allcroft, R., Aspects of aflatoxicosis in farm animals, in *Mycotoxins in Foodstuffs*, M.I.T. Press, Cambridge, Mass., 1965, 153.
3. Allcroft, R. and Carnaghan, R. B. A., Toxic products in groundnuts. Biological effects, *Chem. Ind.* (London), pp. 50-53, 1963.
4. Allcroft, R., Rogers, H., Lewis, G., Nabney, J., and Best, P. E., Metabolism of aflatoxin in sheep: excretion of the "milk toxin", *Nature*, 209, 154, 1966.
5. Anderson, E. E., The nutritive properties of mushrooms *(Agaricus campestris)*, Master's Thesis, Massachusetts State College, 1942.
6. Anderson, E. E. and Fellers, C. R., The food value of mushrooms *(Agaricus campestris), Proc. Amer. Soc. Hort. Sci.*, 41, 301, 1942.
7. Anderson, R. F. and Jackson, R .W., Essential amino acids in microbial proteins, *Appl. Microbiol.*, 6, 369, 1958.
8. Andrews, F. N., Anabolic and estrogenic compound and process of making, U.S. Patent 3,196,019, 1965.
9. Andrews, F. N. and Stob, M., Substance anabolique, Belgian Patent 611,630, 1961.
10. Anonymous, The great truffle snuffle, *Newsweek*, February 24, 1969, pp. 72, 77.
11. Asao, T., Büchi, G., Abdel-Kader, M. M., Chang, S. B., Wick, E. L., and Wogan, G. N., Aflatoxins B and G., *J. Amer. Chem. Soc.*, 85, 1706, 1963.
12. Asao, T., Büchi, G., Abdel-Kader, M. M., Chang, S. B., Wick, E. L., and Wogan, G. N. The structures of aflatoxins B and G_1, *J. Amer. Chem. Soc.*, 87, 882, 163.
13. Asao, T., Büchi, G., Abdel-Kader, M. M., Chang, S. B., Wick, E. L., and Wogan, G. N., The structures of aflatoxins B_1 and G_1, *Mycotoxins in Foodstuffs*. M.I.T. Press, Cambridge, Mass., 1965, 265.
14. Asplin, F. D. and Carnaghan, R. B. A., The toxicity of groundnut meals for poultry with special reference to their effect on ducklings and chickens, *Vet. Rec.*, 73, 1215, 1961.

15. Autret, M. and VanVeen, A. G., Possible sources of proteins for child feeding in underdeveloped countries, *Amer. J. Clin. Nutr.*, 3, 234, 1955.

16. Baker, J. A., Mushroom growing in Province Wellesley and Penang, *Malay Agr. J.*, 22, 25, 1934.

17. Barger, G., *Ergot and Ergotism,* Gurney and Jackson, London, 1931.

18. Bary, A. de, 1858. Ueber die myxomyceten, *Bot. Zeit.*, 16, 357, 361, 356, 1858.

19. Bary, A. de, Die mycetozoen. Ein beitrag zur kenntnis der niedersten thiere, *Zeit. Wiss. Zool.*, 10, 88, 1859.

20. Bary, A. de, *Die Mycetozoen (Schleimpilze). Ein Beitrag zur Kenntnis der niedersten Organismen,* 2nd. ed., W. Engelman, Leipzig, 1864.

21. Bekker, Z. E., *Physiology of Fungi and Its Application in Practice, Moscow University,* (in Russian), 1963.

22. Bell, M. R., Johnson, J. R., Wilde, B. S., and Woodward, R. B., The structure of gliotoxin, *J. Amer. Chem. Soc.*, 80, 1001, 1958.

23. Biollaz, M., Büchi, G., and Milne, G., Biosynthesis of aflatoxin, *J. Amer. Chem. Soc.*, 90, 5017, 1968.

24. Biollaz, M., Büchi, G., and Milne, G., The biosynthesis of bisfuranoids in the genus *Aspergillus, J. Amer. Chem. Soc.*, 90, 5019, 1968.

25. Bird, K., Some economic projections of the U.S. mushroom industry, speech presented at the American Mushroom Institute Conference, Wilmington, Del., September 6, 1969. Reprints available through U.S.D.A. Econ. Res. Service.

26. Block, S. S., Developments in the production of mushroom mycelium in submerged culture, *J. Biochem. Microbiol. Technol. Engineer,* 2, 243, 1960.

27. Block, S. S., Garbage composting for mushroom production, *Appl. Microbiol.*, 13, 5, 1965.

28. Block, S. S., Stearns, T. W., Stephens, R. L., and McCandless, R. F. J., Mushroom mycelium. Experiments with submerged culture, *J. Agr. Food Chem.*, 1, 890, 1953.

29. Block, S. S., Tsao, G., and Han, L., Production of mushrooms from sawdust, *J. Agr. Food Chem.*, 6, 923, 1958.

30. Block, S. S., Tsao, G., and Han, L., Experiments in the cultivation of *Pleurotus ostreatus, Mushroom Sci.*, 4, 309, 1959.

31. Block, S. S., Tsao, G., and Han, L., Experiments in the cultivation of *Pleurotus ostreatus,* Tech. Paper No. 199, Engineering Progress at the University of Florida, Vol. 15, No. 2, 1959.

32. Blount, W. P., Turkey "X" disease, Turkeys 9(2), 52, 55, 61, 77, 1961.

33. Bode, L., La culture artificielle et remuneratrice de la morille. *Petit Casablancais,* 31, 1264, 1946.

34. Boedijn, K. B., Notes on the Mucorales of Indonesia. *Sydowia,* 12, 321, 1958.

35. Borker, E., Insalata, N. F., Levi, C. P., and Witzeman, J. S., Mycotoxins in feeds and foods, *Adv. Appl. Microbiol.*, 8, 315, 1966.

36. Boughey, A. T. and Gray, W. D., The Malala wine industry of Southern Rhodesia, *Proc. 1st Fed. Sci. Congress* (S. Rhodesia), 1, 1960.

37. Brewer, E. N., Kuraishi, S., Garver, J. C., and Strong, F. M., Mass culture of a slime mold, *Physarum polycephalum, Appl. Microbiol.*, 12, 161, 1964.

38. Brock, T. D., Studies on the nutrition of *Morchella esculenta* Fr., *Mycologia,* 43, 402, 1951.

39. Buller, A. H. R., *Researches on Fungi,* Vol. 2, Longmans, Green & Co., London, 1922.

40. Burnside, J. E., Sippel, W. L., Forgacs, J., Carll, W. T., Atwood, M. B., and Doll, E. R., A disease of swine and cattle caused by eating moldy corn. II. Experimental production with pure cultures of molds, *Amer. J. Vet. Res.*, 18, 817, 1957.

41. Camp, W. G., A method of cultivating myxomycete plasmodia, *Bull. Torrey Bot. Club,* 63, 205, 1936.

42. Carll, W. T., Forgacs, J., and Herring, A. S., Toxicity of fungi isolated from a food concentrate, *Amer. J. Hygiene*, 60, 8, 1954.

43. Chahal, D. S. and Gray, W. D., The growth of selected cellulolytic fungi on wood pulp, *Biodeterioration of Materials*, Elsevier Pub. Co., Ltd., London, 1969, 584.

44. Chahal, D. S. and Gray, W. D., 1969b. Growth of cellulolytic fungi on wood pulp. I. Screening of cellulolytic fungi for their growth on wood pulp, *Indian Phytopath*, 22, 79, 1969.

45. Challenger, F., Higginbottom, C., and Ellis L., The formation of organometalloidal compounds by micro-organisms. Part I. Trimethylarsine and dimethylarsine, *J. Chem. Soc.*, pp. 95, 1933.

46. Chang, S. B., Abdel-Kader, M. M., Wick, E. L., and Wogan, G. N., Aflatoxin B_2: chemical identity and biological activity, *Science*, 142, 1191, 1963.

47. Chang, S. T., Pure-culture spawn for *Volvariella volvacea, Chung Chi J.*, 3, 222, 1964.

48. Chang, S. T., The influence of culture methods on the production and nutritive contents of *Volvariella volvacea, Chung Chi J.*, 4, 76, 1964.

49. Chang, S. T., Cultivation of the straw mushroom in S. E. China, *World Crops*, September 1965, pp. 47.

50. Chang, S. T., How to grow straw mushrooms, *Quart. J. Taiwan Museum*, 18, 477, 1965.

51. Chang, S. T., A cytological study of spore germination of *Volvariella volvacea, Bot. Mag. Tokyo*, 82, 102, 1969.

52. Chang, S. T., Factors affecting spore germination in *Volvariella volvacea, Physiol. Plant.*, 22, 734, 1969.

53. Chastukhin, V. Ya., *Mass Cultures of Microscopic Fungi*, Nat. Reservations Headquarters Press, Moscow (in Russian), 1948.

54. Chastukhin, V. Ya., Goncharova, L. A., and Golubchina, R. N., Mass cultures of mycelial fungi as a source of fodder proteins, *Mikrobiologiya*, 26(3), 260, 1957, (Translation published by A.I.B.S.).

55. Chatin, A., *La Truffe*, Paris, 1892.

56. Chavez, M. E., *Liquor: The Servant of Man*. Little, Brown & Co., Boston, 1965.

57. Cheung, K. K. and Sim, G. A., Aflatoxin G_1: direct determination of the structure by the method of isomorphous replacement, *Nature*, 201, 1185, 1964.

58. Christensen, C. M., *Common Fleshy Fungi*, Burgess Pub. Co., Minneapolis, Minn., 1965.

59. Christensen, C. M., Fungi in cereal grains and their products, in *Mycotoxins in Foodstuffs*, M. I. T. Press, Cambridge, Mass., 1965, 9.

60. Christensen, C. M., Nelson, G. H., and Mirocha, C. J., Effect on the white rat uterus of a toxic substance isolated from *Fusarium, Appl. Microbiol.*, 13, 653, 1965.

61. Church, M. B., Laboratory experiments in the manufacture of Chinese ang-khak in the United States. *J. Ind. Eng. Chem.*, 12, 45, 1920.

62. Cirillo, V. P., Crestwood, W. A., Hardwick, O., and Seeley, R. D., Fermentation process for producing edible mushroom mycelium, U.S. Patent 2,928,210, 1960.

63. Clare, N. T., Photosensitization in animals, *Adv. Vet. Sci.*, 2, 182, 1955.

64. Coker, W. C. and Couch J. N., *The Gastromycetes of the Eastern United States and Canada*, University of North Carolina Press, Chapel Hill, N. C., 1928.

65. Costantin, J., La culture de la morille et sa forme conidienne, *Ann. Sci. Nat. Bot.*, 18, 111, 1936.

66. Cottier, C. J., Moore, C. H., Diener, U. L., and Davis, N. D., Effect of continuous feeding of aflatoxin to chickens, *Highlights of Agr. Res.* (Agr. Exp. Sta., Auburn University) 15(3), 7, 1967.

67. Currie, J. N., Flavor of Roquefort cheese, *J. Agr. Res.*, 2, 1, 1914.

68. Daniel, J. W. and Babcock, K., Methionine metabolism of the myxomycete, *Physarum polycephalum, J. Bact.*, 92, 1028, 1966.

69. Daniel, J. W., Babcock, K. L., Sievert, A. H., and Rusch, H. P., Organic requirements and synthetic media for growth of the myxomycete, *Physarum polycephalum, J. Bact.*, 86, 324, 1963.

70. Daniel, J. W., Kelley, J., and Rusch, H. P., Hematin-requiring plasmodial myxomycete, *J. Bact.*, 84, 1104, 1962.

71. Daniel, J. W., and Rusch, H. P., The pure culture of *Physarum polycephalum* on a partially defined soluble medium, *J. Gen. Microbiol*, 25, 47, 1961.

72. Dattilo-Rubbo, S., The taxonomy of fungi of blue-veined cheese, *Trans. Brit. Mycol. Soc.*, 22, 174, 1938.

73. Davis, N. D., Diener, U. L., and Eldridge, D. W., Production of aflatoxins B_1 and G_1 by *Aspergillus flavus* in a semisynthetic medium, *Appl. Microbiol.*, 14, 378, 1966.

74. Davis, N. D., Searcy, J. W., and Diener, U. L., Production of ochratoxin A by *Aspergillus ochraceus* in semisynthetic medium, *Appl. Microbiol.*, 17, 742, 1969.

75. Delmas, J., Orientation des recherches sur la culture du champignon de couche en France, *Mushroom Sci.*, 2, 118, 1954.

76. Diener, U. L. and Davis, N. D., Aflatoxin formation by *Aspergillus flavus*, in *Aflatoxin*, Academic Press, New York, 1969.

77. Diener, U. L., Davis, N. D., Salmon, W. D., and Prickett, C. O., Toxin-producing *Aspergillus* isolated from domestic peanuts, *Science*, 142, 1491, 1963.

78. Djien, K. S. and Hesseltine, C. W., Indonesian fermented foods, *Soybean Digest*, 22, 14, 1961.

79. Doane, C. F. and Lawson, H. W., Varieties of cheese: descriptions and analyses, U.S.D.A. Bull. 608, 1918.

80. Dodd, D. C., Facial eczema in ruminants, in *Mycotoxins in Foodstuffs*, M.I.T. Press, Cambridge, Mass., 1965, 105.

81. Donkersloot, J. A., Hsieh, D.P.H., and Mateles, R. I. Incorporation of precursors into aflatoxin B_1, *J. Amer. Chem. Soc.*, 90, 5020, 1968.

82. Drobotko, V. G., Stachybotryotoxicosis, a new disease of horses and humans, *Amer. Rev. Soviet Med.*, 2, 238, 1945.

83. Drobotko, V. G., Marushenko, P. E., Aizeman, B. E., Kolesnik, N. G., Kudlai, D. B., Iatel, P. D., and Melnichenko, V. D., Stachybotryotoxicosis, a new disease of horses and humans, *Vrach. Delo*, 26, 125, 1946.

84. Duggar, B. M., The cultivation of mushrooms, U.S.D.A. Farmers Bull. No. 204, 1904.

85. Duggar, B. M., Some principles in mushroom growing and mushroom spawn making, *Bull. U.S.D.A.* 85, 1, 1905.

86. Duggar, B. M., *Mushroom Growing*, Orange Judd Co., New York, 1915.

87. Dyson, G. M., Mold food of the Far East, *Pharmaceut. J.*, 121, 375, 1928.

88. Eadie, T. and O'Rear, C. E., The occurrence of aflatoxin in Virginia-North Carolina peanuts and peanut products, *Va. J. Sci.*, 18, 140, 1967.

89. Eddy, B. P., Production of mushroom mycelium by submerged cultivation, *J. Sci. Food Agr.* 9, 644, 1958.

90. Edwards, R. L. and Flegg, P. B., Experiments with artificial mixtures for casing mushroom beds, *Mushroom Sci.*, 2, 143, 1954.

91. Eugster, C. H. and Muller, G., Notiz über weitere Vorkommen von Muscarin, *Helv. Chim. Acta*, 42, 1189, 1959.

92. Falanghe, H., Production of mushroom mycelium as a protein and fat source in submerged culture in medium of vinasse, *Appl. Microbiol.*, 10, 572, 1962.

93. Falanghe, H., Smith, A. K., and Rackis, J. J., Production of fungal mycelial protein in submerged culture of soybean whey, *Appl. Microbiol.*, 12, 330, 1964.

94. Ferguson, M. C., A preliminary study of the germination of spores of *Agaricus campestris* and other basidiomycetous fungi, *Bull. U.S.D.A.*, 16, 1, 1902.

95. Ferreira, N. P., Recent advances in research on ochratoxin. Part 2. Microbiological aspects, in *Biochemistry of Some Foodborne Microbial Toxins.* M.I.T. Press, Cambridge, Mass., 1967, 157.

96. Fitzpatrick, W. H., Esselen, W. B., Jr., and Weir, E., Composition and nutritive value of mushroom protein, *J. Amer. Diet. Ass.,* 22, 318, 1946.

97. Ford, W. W., The distribution of haemolysins, agglutinins and poisons in fungi, especially the Amanitas, the Entolomas, the Lactarius and the Inocybes. *J. Pharmacol. Exp. Ther.,* 2, 285, 1911.

98. Ford, W .W., A new classification of mycetismus (mushroom poisoning), *Trans, Ass. Amer. Physicians,* 38, 225, 1923.

99. Forgacs, J., Mycotoxicoses: The neglected disease, *Feedstuffs,* 34, 124, May 5, 1962.

100. Forgacs, J., Stachybotryotoxicosis and moldy corn toxicosis, in *Mycotoxins in Foodstuffs,* M.I.T. Press, Cambridge, Mass., 1965, 87.

101. Forgacs, J. and Carll, W. T., Mycotoxicoses, *Adv. Vet. Sci.,* 7, 273, 1962.

102. Forgacs, J., Carll, W. T., Herring, A. S., and Henshaw, W. R., Toxicity of *Stachybotrys atra* for animals, *Trans. N.Y. Acad. Sci.,* 20, 787, 1958.

103. Forgacs, J., Carll, W. T., Herring, A. S., and Mahlandt, B. G., A toxic *Aspergillus clavatus* isolated from feed pellets, *Amer. J. Hyg.,* 60, 15, 1954.

104. Fortuskny, V. A., Govrov, A. M., Tebybenko, I. Z., Biochenko, A. S., and Kalitenko, E. T., Stachybotryotoxicosis of cattle and its therapy, *Veterinariya,* 36(9), 67, 1959 (in Russian).

105. Foster, J. W., *Chemical Activities of Fungi,* Academic Press, New York, 1949.

106. Garner, R. J., *Veterinary Toxicology.* Williams & Wilkins, Baltimore, 1961.

107. Genin, A. M. and Shepelev, Y. Y., Certain problems and principles of the formation of a habitable environment based on the circulation of substances, XVth Congress of the Inter. Astronautical Fed., Warsaw, Poland, Sept. 7-12, 1964, Moscow, (in Russian).

108. Gilbert, F. A. and Robinson, R. F., Food from fungi, *Econ. Bot.* 11, 126, 1957.

109. Girolami, R. L. and Knight, S. G., Fatty acid oxidation by *Penicillium roqueforti, Appl. Microbol.,* 3, 264, 1955.

110. Goldblatt, L. A., Ed., *Aflatoxin,* Academic Press, New York, 1969.

111. Golding, N. S., Further studies on the gas requirements of three strains of blue mold *(Penicillium roqueforti* group) isolated from cheese, *J. Dairy Sci.,* 22, 434, 1939.

112. Golding, N. S., The gas requirements of molds. II. The oxygen requirements of *Penicillium roqueforti* (three strains originally isolated from blue veined cheese) in the presence of nitrogen as a diluent and the absence of carbon dioxide, *J. Dairy Sci.,* 23, 879, 1940.

113. Golding, N. S., The gas requirements of molds. III. The effect of various concentrations of carbon dioxide on the growth of *Penicillium roqueforti* (three strains originally isolated from blue-veined cheese) in air, *J. Dairy Sci.,* 23, 891, 1940.

114. Gray, W. D., *The Relation of Fungi to Human Affairs,* Henry Holt and Co., New York, 1959.

115. Gray, W. D., Microbial protein for the space age, *Devel. Ind. Microbiol.,* 3, 63, 1962.

116. Gray, W. D., Fungi as a nutrient source, in *Biologistics for Space Systems Symposium,* 6570th Aerospace Medical Research Laboratories, Wright-Patterson AFB, Ohio. AMRL-TDR-62-116, 1962, p. 356.

117. Gray, W. D., Protein and world population increase, *Proc. 13th Ann. Meet. Agr. Res. Inst.,* p. 97, 1964.

118. Gray, W. D., Process for production of fungal protein, U. S. Patent 3,151,038, 1964.

119. Gray, W. D., Fungi as a potential source of edible protein, Research & Devel. Associates, Activities Report 17, 1, 1965.

120. Gray, W. D., Fungal protein for food and feeds. I. Introduction, *Econ. Bot.,* 20, 89, 1966.

121. Gray, W. D., Fungi and world protein supply, in *World Protein Resources,* Adv. in Chem. Series, 57, 261, 1966.

122. Gray, W. D. and AbouElSoud, M., Fungal protein for food and feeds. II. Whole sweet potato as a substrate, *Econ. Bot.,* 20, 119, 1966.

123. Gray, W. D. and AbouElSeoud, M., Fungal protein for food and feeds. III. Manioc as a potential crude raw material for tropical areas, *Econ. Bot.,* 20, 251, 1966.

124. Gray, W. D. and AbouElSeoud, M., Fungal protein for food and feeds. IV. Whole sugar beets or beet pulp as a substrate, *Econ. Bot.,* 20, 372, 1966.

125. Gray, W. D. and Alexopoulos, C. J., *Biology of the Myxomycetes,* Ronald Press, New York, 1968.

126. Gray, W. D. and Bushnell, W. R., Biosynthetic potentialities of higher fungi, *Mycologia,* 47, 646, 1955.

127. Gray, W. D. and Karve, M. D., Fungal protein for food and feeds. V. Rice as a source of carbohydrate for the production of fungal protein, *Econ. Bot.,* 21, 110, 1967.

128. Gray, W. D. and Paugh, R., Fungal protein for food and feeds. VI. Direct use of cane juice, *Econ. Bot.,* 21, 273, 1967.

129. Gray, W. D. and Staff, I. A., Fungal protein for food and feeds. VII. Caloric values of fungus mycelium, *Econ. Bot.,* 21, 341, 1967.

130. Gray, W. D., Och, F. F., and AbouElSeoud, M., Fungi Imperfecti as potential sources of edible protein, *Devel. Ind. Microbiol.,* 5, 384, 1964.

131. Gray, W. D., Pinto, P. V. C., and Pathak, S. G., Growth of fungi in sea water medium, *Appl. Microbiol.,* 11, 501, 1963.

132. Györky, P., The nutritive value of tempeh, in *Progress in Meeting Protein Needs of Infants and Pre-School Children,* Nat. Acad. Sci., Nat. Res. Council Pub. 843, p. 281, 1961.

133. Hammer, B. W. and Bryant, H. W., A flavor constituent of blue cheese (Roquefort type), *Iowa State Coll. J. Sci.,* 11, 281, 1937.

134. Harshberger, J. W., An ancient Roman toadstool carved in stone, *Mycologia,* 21, 143, 1929.

135. Hartley, R. D., Nesbitt, B. F., and O'Kelley, J., Toxic metabolites of *Aspergillus flavus, Nature,* 198, 1056, 1963.

136. Hebert, A. and Heim, F., Sur la nutrition mineral du champignon de couche, *Ann. Sci. Agron. France et étrangère,* II(III), 1, 1909.

137. Hebert, A. and Heim, F., Nouvelle contribution à l'étude de la nutrition du champignon de couche. Composition des fumiers employés à sa culture, *Ann. Sci. Agron. France et étrangère* III(5), 337, 1911.

138. Heim, R., La culture des morilles, *Rev. Mycol.,* Paris, Suppl. 1, pp. 10, 19, 1936.

139. Heim, R., Les Volvaires, *Rev. Mycol.,* Paris, Suppl. 1, pp. 55, 85, 1936.

140. Henney, H. R., Jr. and Lynch, T., Growth of *Physarum flavicomum* and *Physarum rigidum* in chemically defined minimal media, *J. Bact.,* 99, 531, 1969.

141. Hesseltine, C. W., A millenium of fungi, food, and fermentation, *Mycologia,* 57, 149, 1965.

142. Hesseltine, C. W. and Shibasaki, K., Miso III. Pure culture fermentation with *Saccharomyces rouxii, Appl. Microbiol.,* 9, 515, 1961.

143. Hesseltine, C. W. and Wang, H. L., Traditional fermented foods, *Biotech, Bioengineer,* 9, 275, 1967.

144. Hesseltine, C. W., Shotwell, O. L., Ellis, J. J., and Stubblefield, R. D., Aflatoxin formation by *Aspergillus flavus, Bact. Rev.,* 30, 795, 1966.

145. Hiscocks, E. S., The importance of molds in the deterioration of tropical foods and feedstuffs, in *Mycotoxins in Foodstuffs.* M.I.T. Press, Cambridge, Mass., 1965, 15.

146. Holker, J. S. E. and Underwood, J. G., A synthesis of cyclopentenocoumarin structurally related to aflatoxin B, *Chem. Ind.* (London), 3, 1865, 1964.

147. Houck, J. P., The Pennsylvania mushroom industry—problems and prospects, *Farm Economics*, February 1, 1964. Extension Service, U.S.D.A., The Pennsylvania State University.

148. Howard, F. L., Laboratory cultivation of myxomycete plasmodia, *Amer. J. Bot.*, 18, 624, 1931.

149. Humfeld, H., The production of mushroom mycelium *(Agaricus campestris)* in submerged culture, *Science*, 107, 373, 1948.

150. Humfeld, H., Production of mushroom mycelium, U.S. Patent 2,618,900, 1952.

151. Humfeld, H., Production of mushroom mycelium by submerged culture in liquid medium, U.S. Patent 2,693,665, 1954.

152. Humfeld, H. and Sugihara, T. F., Mushroom mycelium production by submerged propagation, *Food Technol.*, 3, 355, 1949.

153. Humfeld, H. and Sugihara, T. F., The nutrient requirements of *Agaricus campestris* grown in submerged culture, *Mycologia*, 44, 605, 1952.

154. Hutchison, H. B. and Richards, E. H., Artificial farmyard manure, *J. Min. Agr. Great Brit.*, 28, 398, 1921-23.

155. Iizuka, H. and Iida, M., Maltoryzine, a new toxic metabolite produced by a strain of *A. oryzae* var. *microsporus* isolated from the poisonous malt sprout, *Nature*, 196, 681, 1962.

156. Jackson, E. W., Wolf, H., and Sinnhuber, R. O., The relationship of hepatoma in rainbow trout to aflatoxin contamination and cottonseed meal, *Cancer Res.*, 28, 987, 1968.

157. Jenkins, S. H., Organic manures, *Tech. Com. Imp. Bur. Soil Sci.*, 33, 1935.

158. Jennison, M. W., Newcomb, M. D., and Henderson, R., Physiology of the wood-rotting Basidiomycetes. I. Growth and nutrition in submerged culture in synthetic media, *Mycologia*, 47, 275, 1955.

159. Jennison, M. W., Richberg, C. G., and Krikszens, A. E., Physiology of wood-rotting Basidiomycetes. II. Nutritive composition of mycelium grown in submerged culture, *Appl. Microbiol.*, 5, 87, 1957.

160. Jensen, O., Biologische Studien über den Kaserifungsprozesc unter spezieller Berücksichtigung den flüchtigen Fettsäuren, in *Landw. Jahrb. Schweiz.*, Jahrg. 18, Heft 8, 1904, 319.

161. Joffee, A. Z., Toxicity of overwintered cereals, *Plant Soil*, 18(1), 31, 1963.

162. Joffee, A. Z., Toxin production by cereal fungi causing toxic alimentary aleukia in man, in *Mycotoxins in Foodstuffs*, M.I.T. Press, Cambridge, Mass., 1965, 77.

163. Johnson, J. R., and Buchanan, J. B., Gliotoxin. X. Dethiogliotoxin and related compounds, *J. Amer. Chem.* Soc., 75, 2103, 1953.

164. Johnson, J. R., Bruce, W. F., and Dutcher, J. D., Gliotoxin, the antibiotic principle of *Gliocladium fimbriatum*. I. Production, physical and biological properties, *J. Amer. Chem. Soc.*, 65, 2005, 1943.

165. Kelley, A. P., *Mycotrophy in Plants*, Chronica Botanica Co., Waltham, Mass., 1950.

166. Kelley, J., Daniel, J. W., and Rusch, H. P., A hemin-requiring plasmodial slime mold, *Fed. Proc.*, 19, 243, 1960.

167. Kinosita, R. and Shikata, R., On toxic moldy rice, in *Mycotoxins in Foodstuffs*, M. I. T. Press, Cambridge, Mass., 1965, 111.

168. Kirby, G. W., Frey, C. N., and Atkin, L., The growth of bread molds as influenced by acidity, *Cereal Chem.*, 12, 244, 1935.

169. Kirby, G. W., Frey, C. N., and Atkin, L., Further studies on the growth of bread mold as influenced by acidity, *Cereal Chem.*, 14, 865, 1937.

170. Klebs, G., Zur Physiologie der Fortpflangung einiger Pilze III. Allgemeine Betractungen, *Jahrb, wiss. Bot.*, bd. 35, Heft 1. 80, 1900.

171. Kneebone, L. R., Improvements in strain provide better spawn, *Pennsylvania Packer*, April-May-June 1969, 7.

172. Knight, S. G., Mohr, W. H., and Frazier, W. C., White mutants of *Penicillium roqueforti*, *J. Dairy Sci.*, 33, 929, 1950.

173. Kobayashi, Y., Uraguchi, K., Sakai, F., Tatsuno, T., Tsukioka, M., Sakai, Y., Sato, T., Miyake, M., Enomoto, M., Shikata, T., and Ishiko, T., Toxicological studies on the yellowed rice by *P. isl.* Sopp. I. Experimental approach to liver injuries by long term feedings with the noxious fungus on mice and rats, *Proc. Jap. Acad.*, 34, 139, 1958.

174. Kobayashi, Y., Uraguchi, K., Sakai, F., Tatsuno, T., Tsukioka, M., Sakai, Y., Sato, T. Miyake, M., Enomoto M., Shikata, T., and Ishiko, T., Toxicological studies on the yellowed rice by *P. isl.* Sopp. II. Isolation of the two toxic substances from the noxious fungus, and their chemical and biological properties, *Proc. Jap. Acad.*, 34, 736, 1959.

175. Kobayashi, Y., Uraguchi, K., Sakai, F., Tatsuno, T., Tsukioka, M., Sakai, Y., Sato, T., Miyake, M., Enomoto, M., Shikata, T., and Ishiko, T., Toxicological studies on the yellowed rice by *P. islandicum*. III. Experimental verification of primary hepatic carcinoma of rats by long-term feeding with the fungus-growing rice, *Proc. Jap. Acad.*, 34, 501, 1959.

176. Kondō M. and Kashara, Y., Versuche bezüglich der Aufbewahrung der Sporen von Shiitake, *Cortinellus shiitake* Schröt, *Ber. Ohara Inst.*, 6, 28, 1933.

177. Kraybill, H. F. and Shapiro, R. E., Implications of fungal toxicity to human health, in *Aflatoxins*, Academic Press, New York, 1969, 401.

178. Kraybill, H. F. and Shimkin, M. B., Carcinogenesis related to foods contaminated by processing and fungal metabolites, *Adv. Cancer Res.*, 8, 191, 1964.

179. Kuninaka, A., Kibi, M., and Sakaguchi, K., History & development of flavor nucleotides, *Food Technol.*, 18, 287, 1964.

180. Lambert, E. B., Studies on the relation of temperature to the growth, parasitism, thermal death points and control of *Mycogone perniciosa*, *Phytopathology*, 20, 75, 1930.

181. Lambert, E. B., Effect of excess carbon dioxide on growing mushrooms, *J. Agr. Res.*, 47, 599, 1933.

182. Lambert, E. B., Principles and problems of mushroom culture, *Bot. Rev.*, 4, 397, 1938.

183. Lambert, E. B., Mushroom casing soil in relation to yield, *Dept. Circ. U.S.D.A.*, 509, 1, 1939.

184. Lambert, E. B., Indoor composting for mushroom culture, *Dept. Circ. U.S.D.A.*, 609, 1, 1941.

185. Lambert, E. B., and Ayers, T. T., An improved system of mushroom culture for better control of diseases, *Plant Dis. Reptr.*, 36, 261, 1952.

186. Lancaster, M. C., Jenkins, F. P., and Phelp, J., Toxicity associated with certain samples of groundnuts, *Nature*, 192, 1095, 1961.

187. Langkramer, C. and Rězník, A., Umělá mrva-náhradní substrát pro pěstováni žampionů, Čes. *Mikoligie*, 8, 172, 1954 (*c. f.* Singer, 1961).

188. Liener, I. E., Toxic substances associated with seed proteins, in *World Protein Resources*, Adv. in Chem. Series, 57, 178, 1966.

189. Liese, J., Ueber die Moeglichkeit einer Pilzzucht in Walde, *Der Deutsche Forstbeamte*, No. 25, 1934.

190. Lintzel, W., Über den Nahrwert den Einweisses der Speisepilze, *Biochem. Zeit.*, 308, 413, 1941.

191. Litchfield, J. H. and Overbeck, R. C., Submerged culture growth of *Morchella* species in food processing waste substrates, *Proc. 1st Intern. Cong. Food Sci. and Technol.*, London, Gordon and Breech, New York, B-6, Part II, 1963.

192. Litchfield, J. H., Overbeck, R. C., and Davidson, R. S., Factors affecting the growth of morel mushroom mycelium in submerged culture, *J. Agr. Food Chem.* 11, 158, 1963.

193. Litchfield, J. H., Vely, V. G., and Overbeck, R. C., Nutrient content of morel mycelium: amino acid composition of the protein, *J. Food Sci.*, 28, 741, 1963.

194. Lockwood, L. B., The production of Chinese soya sauce, *Soybean Digest*, 7, 10, 1947.

195. Lockwood, L. B. and Smith, A. K., Fermented soy foods and sauce, Yearbook Separate No. 2213, from pp. 357-361 of the 1950-51 Yearbook of Agriculture.

196. Lohwag, H., Über Trüffelvorkommen, *Verh. Zool.-Bot. Ges. Wien.*, 82, 117, 1932.

197. Loosmore, R. M. and Markson, L. M., Poisoning of cattle by Brazilian groundnut meal, *Vet. Rec.*, 73, 813, 1961.

198. Madhavan, T. V. and Gopalan, C., Effect of dietary protein on aflatoxin liver injury in weanling rats, *Arch. Path.*, 80, 123, 1965.

199. Madhavan, T. V., Rao, K. S., and Tulpule, P. G., Effect of dietary protein on levels of susceptibility of monkeys to aflatoxin liver injury, *Indian J. Med. Res.*, 53, 984, 1965.

200. Majumder, S. K., Narasimhan, K. S., and Parpia, H. A. B., Microecological factors of microbial spoilage and the occurrence of mycotoxins on stored grains, in *Mycotoxins in Foodstuffs*, M.I.T. Press, Cambridge, Mass., 1965, 27.

201. Malençon, G., Les truffes europeénes, historique, morphogénie, organographie, classification, culture, *Rev. Mycol.*, Paris (Mém. hors serie), 1, 1938.

202. Martin, G. W., The numbers of fungi, *Proc. Iowa Acad. Sci.*, 58, 175, 1951.

203. Martinelli, A. F. and Hesseltine, C. W., Tempeh fermentation: package and tray fermentations, *Food Technol.*, 18, 167, 1964.

204. Marumo, S., Islanditoxin, a toxic metabolite produced by *Penicillium islandicum*. III. Structure of islanditoxin, *Bull. Agr. Chem. Soc. Jap.*, 23, 428, 1959.

205. Mateles, R. I. and Wogan, G. N., Eds., *Biochemistry of Some Foodborne Microbial Toxins*. M.I.T. Press, Cambridge, Mass., 1967.

206. Matruchot, L., Germination des spores de truffes; culture et caractère du mycélium truffier, *C. R. Acad. Sci.*, Paris, 124, 1099, 1903.

207. Mayer, C. F., Endemic panmyelotoxicosis in the Russian grain belt. Part I. The clinical aspects of alimentary toxic aleukia (ATA), *Military Surgeon*, 113, 173, 1953. Part II. The botany, phytopathology and toxicology of Russian cereal feed, *Military Surgeon*, 113, 295, 1953.

208. McConnell, J. E. W. and Esselen, W. B., Jr., Carbohydrates in cultivated mushrooms *(Agaricus campestris)*, *Food Res.*, 12, 118, 1947.

209. McNutt, S. H., Purwin, P., and Murray, C., Vulvovaginitis in swine, Preliminary report, *J. Amer. Vet. Med. Ass.*, 73, 484, 1928.

210. Milner, M., Mycotoxins and protein food efforts, in *Mycotoxins in Foodstuffs*, M.I.T. Press, Cambridge, Mass., 1965, 69.

211. Mirocha, C. J., Christensen, C. M., and Nelson, G. H., An estrogenic metabolite produced by *Fusarium graminearum* in stored corn, in *Biochemistry of Some Foodborne Microbial Toxins*, M.I.T. Press, Cambridge, Mass., 1967, 119.

212. Miyake, M. and Saito, M., Liver injury and liver tumors induced by toxins of *Penicillium islandicum* Sopp growing on yellowed rice, in *Mycotoxins in Foodstuffs*, M.I.T. Press, Cambridge, Mass., 1965, 133.

213. Molliard, M., Recherches calorimétriques sur l'utilization de l'énergie respiratorie au cours du développement d'une culture de *Sterigmatocystis nigra*, *Compt. Rend. Soc. Biol.*, 87, 219, 1922.

214. Morris, H. A., Jezeski, J. J., and Combs, S. T., The use of white mutants of *Penicillium roqueforti* in cheese making, *J. Dairy Sci.*, 35, 480, 1952.

215. Morris, H. A., Jezeski, J. J., and Combs, S. T., The use of white mutants of *Penicillium roqueforti* in cheese making, *J. Dairy Sci.*, 37, 711, 1954.

216. Moustafa, A. M., Nutrition and development of mushroom flavor in *Agaricus campestris* mycelium, *Appl. Microbiol.*, 8, 63, 1960.

217. Nakano, M., Traditional methods of food processing. Regional Seminar on Food Technology for Asia and the Far East. Food Res. Inst., Ministry of Agriculture and Forestry, Tokyo, Japan, p. 1, 1959.

218. Nesbitt, B. F., O'Kelley, J., Sargeant, K., and Sheridan, A., Toxic metabolites of *Aspergillus flavus*, *Nature*, 195, 1062, 1962.

219. Nishikawa, H., Biochemistry of filamentous fungi. I. Coloring matters of *Monascus purpureus* Went, *Agr. Chem. Soc. Jap.*, 8, 1007, 1932.

220. Nisikado, Y. and Yamauti, K., Studies on the heterothallism of *Cortinellus berkeleyana* Ito & Imai, an economically important edible mushroom in Japan, *Ber. Ohara Inst.*, 7, 115, 1935.

221. Ochse, J. J., *Vegetables of the Dutch East Indies,* Archipel. Drukkeriji, Buitenzorg, Java, 1931.

222. Oettle, A. G., Cancer in Africa, especially in regions south of the Sahara, *J. Nat. Cancer Inst.*, 33, 383, 1964.

223. Olifson, L. E., Toxins isolated from overwintered cereals and their chemical nature. Monitor, Orenburg Sect. of the U.S.S.R., *D. J. Mendeleyev Chem. Soc.*, 9, 21, 1957. (in Russian).

224. Onishi, H., Studies on osmophilic yeasts, *Bull. Agr. Chem. Soc. Jap.*, 21, 137, 143, 151; 23, 332, 351, 359; 24, 126, 131, 226, 386, *Agr. Biol. Chem.*, 25, 341, 1957-61.

225. Palo, M. A., Vidal-Adeva, L., and Maceda, L., A study of ang-khak and its production, *Philippine J. Sci.*, 89, 1, 1961.

226. Parrish, F. W., Wiley, B. J., Simmons, E. G., and Long, L., Jr., Production of aflatoxins and kojic acid by species of *Aspergillus* and *Penicillium, Appl. Microbiol,* 14, 139, 1966

227. Passecker, F., Kulturversuche mit dem japanischen Shiitake oder Pananiapilz, *Gartenbauwiss,* 8, 359, 1933.

228. Pathak, S. G. and Seshadri, R., Use of *Penicillium chrysogenum* as animal food, *Appl. Microbiol.*, 13, 262, 1965.

229. Patton, S., The methyl ketones of blue cheese and their relation to flavor, *J. Dairy Sci.*, 33, 680, 1950.

230. Platt, B. S., Tables of representative values of foods commonly used in tropical countries. Great Britain Medical Research Council, Special Report, Ser. 23, p. 12, 1945.

231. Prescott, S. C. and Dunn, C. G., *Industrial Microbiology,* 2nd ed., McGraw-Hill Book Co., New York, 1949.

232. Pringsheim, H. and Lichtenstein, S., Versuche zur Anreicherung von Kraftstroh mit Pilzeiweiss, *Cellulosechemie,* 1, 29, 1920.

233. Purchase, I. F. H. and Nel, W., Recent advances in research on ochratoxin. Part 1. Toxicological aspects, in *Biochemistry of Some Foodborne Microbial Toxins,* M.I.T. Press, Cambridge, Mass., 1967, 153.

234. Rainbow, C., and Rose, A. H., *Biochemistry of Industrial Microorganisms,* Academic Press, New York, 1963.

235. Raistrick, H., Birkinshaw, J. H., Charles, H. H. V., Clutterbuck, P. W., Coyne, F. P., Hetherington, A. C., Lilly, C. H., Rintoul, M. L., Rintoul, W., Robinson, R., Stoyle, J. A. R., Thom, C., and Young, W., Studies in the biochemistry of microorganisms, *Phil. Trans. Roy. Soc. London,* Ser. B. 220, 1, 1931.

236. Ramsbottom, J., *Mushooms & Toadstools,* Collins, St. James Place, London, 1953.

237. Rautavaara, T., Suomen Sienstato. Porwoo, Helsinki, 1947 (*c.f.* Singer, 1954).

238. Rettew, G. R., Mushroom spawn and substrate therefor, U.S. Patent 1,939,600, 1933.

239. Rettew, G. R. and Thompson, F. G., *Manual of Mushroom Culture,* Mushroom Supply Co., Toughkenamon, Pennsylvania, 1948.

240. Reusser, F., Spencer, J. F. T., and Sallans, H. R., Protein and fat content of some mushrooms grown in submerged culture, *Appl. Microbiol.*, 6, 1, 1958.

241. Reusser, F., Spencer, J. F. T., and Sallans, H. R. *Tricholoma nudum* as a source of microbiological protein, *Appl. Microbiol.*, 6, 5, 1958.

242. Robinson, R. F. and Davidson, R. S., The large-scale growth of higher fungi, *Adv. Appl. Microbiol.*, 1, 261, 1959.

243. Rock, J., *The Time Has Come,* Alfred A. Knopf, New York, 1963.

244. Rolfe, R. T. and Rolfe, F. W., *The Romance of the Fungus World,* Lipppincott,, Philadelphia, 1928.

245. Rollofsen, P. A. and Talens, A., Changes in some B-vitamins during molding of soybeans by *Rhizopus oryzae* in the production of tempeh kedelee, *J. Food Sci.*, 29, 224, 1964.

246. Rothlin, E. and Bircher, R., Allergy, the autonomic nervous system and ergot alkaloids, *Progr. Allerg.*, 3, 434, 1952.

247. Rubner, M., Energieverbrauch im Leben der Mikroorganismen, *Arch. Hyg.*, 48, 260, 1904.

248. Saisithi, P., Kasemsarn, B., Liston, J., and Dollar, A. M., Microbiology and chemistry of fermented fish, *J. Food Sci.*, 31 105, 1966.

249. Sallet, A., La culture d'un champignon d'Annam: le Nâm Rom 'champignon des pailles', *Rev. Mycol.*, Paris, Suppl. 1(6), 91, 1936.

250. Sargeant, K., Sheridan, A., O'Kelley, J., and Carnaghan, R. B. A., Toxicity associated with certain samples of groundnuts, *Nature*, 192, 1096, 1961.

251. Sarkisov, A. Ch., Mycotoxicoses, Govt. Edit. Agr. Lit., Moscow, 1954 (in Russian).

252. Savage, G. M., and VanderBrook, M. J., The fragmentation of the mycelium of *Penicillium notatum* and *Penicillium chrysogenum* by a high-speed blender and the evaluation of the blended seed, *J. Bact.*, 52, 385, 1946.

253. Schroeder, H. W. and Hein, H., Aflatoxins: production of the toxins in vitro in relation to temperature, *Appl. Microbiol.*, 15, 441, 1967.

254. Scott, de B., Toxigenic fungi isolated from cereal and legume products, Mycopathologia, 14, 213, 1965.

255. Scott, R., Blue veined cheese, *Proc. Biochem.*, 3, 11, 1968.

256. Searcy, J. W., Davis, N. D. and Diener, U. L., Biosynthesis of ochratoxin A, *Appl. Microbiol*, 18, 622, 1969.

257. Shibasaki, K. and Hesseltine, C. W., Miso I. Preparation of soybeans for fermentation, *J. Biochem. Microbiol. Technol.*, 3, 161, 1961.

258. Shibasaki, K. and Hesseltine, C. W., Miso II. Fermentation, *Devel. Ind. Microbiol.*, 2, 205, 1961.

259. Shibasaki, K. and Hesseltine, C. W., Miso fermentation, *Econ. Bot.*, 16, 180, 1962.

260. Shibata, S. and Kitagawa, I., Metabolic products of fungi. X. Structure of rubroskyrin and its relation to the structure of luteoskyrin, *Pharm. Bull.* (Tokyo), 4, 309, 1956.

261. Shibata, S. and Kitagawa, I., Metabolic products of fungi. Structures of rubroskyrin and luteoskyrin, *Chem. Pharm. Bull.* (Tokyo), 8, 884, 1960.

262. Shotwell, O. L., Hesseltine, C. W.., Burmeister, H. R., Kwolek, W. F., Shannon, G. M., and Hall, H. H., Survey of cereal grains and soybeans for the presence of aflatoxin. I. Wheat, grain sorghum and oats, *Cereal Chem.*, 47, 446. II. Corn and soybeans, *Cereal Chem.*, 47, 454, 1969.

263. Sinden, J. W. and Hauser, E., The short method of composting, *Mushroom Sci.*, 1, 52, 1950.

264. Singer, R., Is Shiitake a *Cortinellus? Mycologia*, 33, 449, 1941.

265. Singer, R., Cryptogamic flora of the Arctic. VI. Fungi, *Econ. Bot.*, 20, 451, 1954.

266. Singer, R., *Mushrooms and Truffles*, Leonard Hill (Books) Ltd., London, 1961.

267. Sinnhuber, R. O., Wales, J. H., Engebrecht, R. H., Amend, D. F., Kray, W. D., Ayers, J. L., and Ashton, W. E., Aflatoxins in cottonseed meal and hepatoma in trout, *Fed. Proc.*, 24, (2, pt. I), 627, 1965.

268. Sipple, W. L., Mold intoxication of livestock, *Iowa Vet.*, 28, 15, 42, 1957.

269. Sipple, W. L., Burnside, J. E., and Atwood, M. B., A disease of swine and cattle caused by eating moldy corn, Proc. 90th Ann. Meeting Amer. Vet. Med. Ass., Toronto, Canada, p. 174, 1953.

270. Skinner, C. S., The synthesis of aromatic amino acids from inorganic nitrogen by molds and the value of mold proteins in the diet, *J. Bact.*, 28, 95, 1924.

271. Skinner, J. T., Peterson, W. H., and Steenbock, H., Nährwert von Schimmelpilzmycel, *Biochem. Zeit.*, 267, 169, 1933.

272. Smith, A. H., *Mushrooms in Their Natural Habitats*. Vol. I., Sawyers, Inc., Portland, Oreg., 1949.

273. Smith, A. H., *The Mushroom Hunter's Field Guide,* University of Michigan Press, Ann Arbor, Mich. 1963.

274. Smith, A. K., Foreign uses of soybean protein foods, *Cereal Sci. Today*, 8, 196, 198, 200, 210, 1963.

275. Smith, A. K., Hesseltine, C. W., and Shibasaki, K., Preparation of miso, U.S. Patent 2,967,108, 1961.

276. Smith, A. K., Rackis, J. J., Hesseltine, C. W., Smith, M., Robbins, D. J., and Booth, A. N., Tempeh: nutritive value in relation to processing, *Cereal Chem.*, 41, 173, 1964.

277. Smith, A. K., Watanabe, T., and Nash, A. M., Tofu from Japanese and United States soybeans, *Food. Technol.*, 14, 332, 1960.

278. Sorenson, W. G., and Hesseltine, C. W., Carbon and nitrogen utilization by *Rhizopus oligosporus, Mycologia*, 58, 681, 1966.

279. Spensley, P. C., Aflatoxin, the active principle of turkey "X" disease, *Endeavour*, 22, 75, 1963.

280. Spoelstra, P. A., Air-conditioning of mushroom houses, *Mushroom Sci.*, 2, 112, 1954.

281. Staley, A. R., Soy sauce goes American, *Food Industries*, 7, 66, 1925.

282. Stanton, W. R. and Wallbridge, A., Fermented food processes, *Proc. Biochem.*, 4, 45, 1969.

283. Steinkraus, K. H., Hand, D. B., Van Buren, J. P., and Hackler, L. R., Pilot plant studies on tempeh, Proc. Conf. Soybean Products for Protein in Human Foods, U.S.D.A., Ars-17-22, p, 83, 1961.

284. Steinkraus, K. H., Lee, C. Y., and Buck, P. A., Soybean fermentation by the ontjom mold *Neurospora. Food Technol.*, 19, 1301, 1965.

285. Steinkraus, K. H., Yap Bwee Hwa, Van Buren, J. P., Provvidenti, M. I., and Hand, D. B., Studies on tempeh—an Indonesian fermented soybean food, *Food. Res.*, 25, 777, 1960.

286. Stob, M., Baldwin, R. S., Tuite, J., Andrews, F. N., and Gillette, K. G., Isolation of an anabolic, uterotropic compound from corn infected with *Gibberella zeae, Nature*, 196, 1318, 1962.

287. Stokes, J. L., and Gunness, M., The effect of cultural conditions on the amino acid content of *Penicillium notatum, J. Bact.*, 52, 195, 1946.

288. Stoller, B. B., Principles and practice of mushroom culture, *Econ. Bot.*, 8, 48, 1954.

289. Stoller, B. B. and Stauffer, J. F., Studies on naturally occurring and ultraviolate radiation induced strains of the cultivated mushroom *Agaricus campestris* L., *Mushroom Sci.*, 2, 51, 1964.

290. Styer, J. F., Nutrition of the cultivated mushroom, *Amer. J. Bot.*, 17, 982, 1930.

291. Su, U Thet, Report of the Mycologist Burma Mandalay for the Year Ending 31st of March 1936, 4, Mandalay, 1936 (*c. f.* Singer, 1961).

292. Sugihara, T. F., and Humfeld, H., Submerged culture of the mycelium of various species of mushroom, *Appl. Microbiol.*, 2, 170, 1954.

293. Szuecs, J., Essence of mushroom and its preparation, U.S. Patent 2,508,811, 1950.

294. Szuecs, J., Method of enhancing mushroom mycelium flavor, U.S. Patent 2,693,664, 1954.

295. Szuecs, J., Mushroom culture, U.S. Patent 2,761,246, 1956.

296. Szuecs, J., Method of growing mushroom mycelium and the resulting products, U.S. Patent 2,850,841, 1958.

297. Taber, R. A. and Schroeder, H. W., Aflatoxin-producing potential of isolates of the *Aspergillus flavus-oryzae* group from peanuts *(Arachnis hypogaea), Appl. Microbiol.*, 15, 140, 1967.

298. Takata, R., The utilization of microorganisms for human food materials, *J. Soc. Chem. Ind. Japan*, 32, 243, 1929.

299. Tamiya, H., Zur Energetik des Wachstums. II. Beitrag zur Atmungsphysiologie der Schimmelpilze, *Acta Phytochim.*, 6, 265, 1932.

For Product Safety Concerns and Information please contact our EU
representative GPSR@taylorandfrancis.com
Taylor & Francis Verlag GmbH, Kaufingerstraße 24, 80331 München, Germany

www.ingramcontent.com/pod-product-compliance
Ingram Content Group UK Ltd.
Pitfield, Milton Keynes, MK11 3LW, UK
UKHW052222190926
475702UK00006B/1154
* 9 7 8 1 3 1 5 8 9 8 3 7 7 *